SO YOU WANT TO BE AN
ENGINEER?

A Guide to Success in the Engineering Profession

Frederick Fell Publishers, Inc
2131 Hollywood Blvd., Suite 305
Hollywood, Fl 33020
www.Fellpub.com
email: Fellpub@aol.com

Frederick Fell Publishers, Inc
2131 Hollywood Blvd., Suite 305
Hollywood, Fl 33020

For information about special discounts for bulk purchases, Please contact Frederick Fell Special Sales at business@fellpub.com.

Designed by Elena Solis

Manufactured in the United States of America

10 9 8 7 6 5 4 3 2 1

Library of Congress Cataloging-in-Publication Data

Calabrese, Marianne Pilgrim, 1947-
 So you want to be an engineer : a guide to success in the engineering profession / by Marianne Pilgrim Calabrese & Ron Davidson.
     p. cm.
Includes bibliographical references.
ISBN 0-88391-187-6 (pbk.)
1. Engineering--Vocational guidance. I. Davidson, Ron, 1945- II. Title.
TA157.C253 2008
620.0023--dc22

                              2008042749

ISBN 13: 978-0-88391-187-7

SO YOU WANT TO BE AN
ENGINEER?

A Guide to Success in the
Engineering Profession

Marianne Pilgrim Calabrese & Ron Davidson

Dedication

This book is dedicated to all the nerds of the world. We know you've been misunderstood and in some cases ridiculed, but without you, the world would not be the wondrous place we live in. Imagine a world without roads, bridges, tunnels, automobiles or airplanes. Imagine a world without radio and television, newspapers or the internet. Imagine a world without anesthesia, antibiotics or vaccines.

Thank you all for wanting to become Engineers.

Acknowledgments

I would like to give a special acknowledgement to Robert E. Albano for his immense help, expert advice and support in making this book possible. Also special thanks to David James Munz for his modern and up to date advice on the engineering profession and to all the engineers that helped us with this book including Glenda S Cass.

To our friends and family who gave us love and support, Susanne and John Calabrese, Ann and Dan Davidson, Leslie and Stu Israel, Theresa Munz, and Francine Pilgrim.

Authors Biography

Ron Davidson

Ron Davidson earned his Bachelors and Masters Degree in Electrical Engineering at New York University in the late 1960's. He began his career in New York designing computers when they were still refrigerator size. Early in the 1970's, he went back to school for his MBA and soon thereafter moved to Silicon Valley, CA where he worked for a startup semiconductor company. There, he had the opportunity to become a Sales Engineer and remained in sales for the rest of his career. Over the years he has sold products from the smallest integrated circuits to the largest networking systems. Ron has accumulated a vast amount of information to help engineers in all situations and in all areas of engineering. He remains an electrical tinker to this day and resides in Wellington, Florida.

Marianne Pilgrim Calabrese

Marianne Pilgrim Calabrese earned her Masters Degree in Reading from Adelphi University and her Bachelor Degree from Queens College in education and psychology. She is a multi published author, teacher, and renowned artist. She has extensive teaching experience and received "The Presidents Club Award" from the International Reading association for her work. Marianne has participated in educational councils and comprehensive teaching in both public and private schools. She also volunteered and has done extensive research for her writing. At present she is a teacher, writer and artist working and living in Wellington Florida. With this knowledge and experience she has accumulated helpful information for her readers.

Scope, Plan and Purpose

So You Want to Be An Engineer? Is a book for anyone who is or who wants to be an Engineer. The book reveals everything no one else will tell you about the engineering profession. It shows how to save you the agony of on the job trial and error training and will give you a head start in using experienced strategies while dealing with technicians, draftsman, marketing, purchasing and manufacturing personnel, project managers and clients as well as other engineers. It doesn't teach you about engineering; it enlightens you about the different aspects of an engineering career. It will tell you what type of engineering will be best for you and where to find your right position. There are The Ten Commandments for an engineer, which sums up in ten steps how to survive in the engineering profession and gives in depth reasons why they work.

It is a refreshingly new and realistic book that touches on the reality that engineers may succeed, not only because of their technical expertise but because of the way they interact with their colleagues, technicians, draftsman, marketing, purchasing and manufacturing personnel, project managers and senior manages of the organization. The chapter topics are;

1. Why do you want to be an engineer?
2. Types of Engineering
3. Choosing the right engineering field for you
4. Education and Licensing
5. Salaries and Positions
6. Meet the engineers
7. Team Work and Interdepartmental relationships with other engineers
8. Meetings and Conferences
9. How to make your company work for you
10. The Ten Commandments of an Engineer

Each of these topics will be discussed fully with real life stories and examples. There will be easy steps given on how to handle each issue and how an engineer can ease into a firm. The Ten Commandments will make it easy for you to sum up the do's and don'ts to survive in the engineering profession.

Preface

So You Want To Be An Engineer? Conveys a realistic way of looking at an engineer's role. It tells you what no one else will tell you about the engineering profession. Entertaining narrative stories are incorporated into the text to give the reader examples of real life situations. These authentic tales will show how engineers behave and react to technicians, draftsman, marketing, purchasing and manufacturing personnel, and project managers and how one can best handle each situation.

The Ten Commandments of an Engineer are provided as a guide to success in easing into any engineering situation and also serves as a summary of the main ideas from the book. Anyone interested in a career in the engineering profession and how to survive in it will find this book both worthwhile and entertaining reading. It is for anyone who wants to be an engineer, who is an engineer, has just started their career in engineering, or is in engineering school. Anyone who is considering becoming an engineer or contemplating a change of careers will also find the facts as expressed in, So You Want To Be An Engineer? very influential in their decision making. This book will give everyone a new outlook on life in the engineering profession.

Table of Contents

Introduction

So You Want to Be An Engineer? Shows you ways to ease into any engineering situation with coworkers, clients, and managers. It also gives you helpful hints through the Ten Commandments of Engineering on how to find the best types of engineering for you, colleges and licenses, and company polices.

CHAPTER 1: WHY DO YOU WANT TO BE AN ENGINEER?
* Personality Quiz to get to know your personality
* Career Choice process
* What do you expect from the engineering profession
* Definition of an engineer

This chapter will help you decide if you want to be an engineer and the many reasons for choosing the engineering profession. If you love math and science and have always been fascinated with tinkering you now know you are heading toward the right profession.

You will examine your personality and get insight into determining whether the engineering profession is for you. You will reviewed the two career choice processes involved, self analysis and personal goals and you will search your natural professional abilities to determine your potential for becoming an engineer.

We will delve into what you should expect from the engineering profession and self exploratory questions will help you have a clear and focused career expectation.

The definition of an engineer, from the American Society of Engineering Education, will give you a broad idea of the engineering profession.

CHAPTER 2: TYPES OF ENGINEERING
* Engineering career paths
* Traditional and Specialty Areas of The Profession
* Chemical Engineers, Civil Engineers, Electrical & Electronics

Engineers, Industrial Engineers, Materials Science Engineers, and Mechanical Engineers
- Aerospace Engineers, Agricultural Engineers, Automotive Engineers, Biomedical Engineers, Ceramics Engineers, Computer Engineers, Environmental Engineers, Geological Engineers, Manufacturing Engineers, Metallurgical Engineers, Mining Engineers, Nuclear Engineers, Optical engineers, Packaging Engineers, Petroleum Engineers, Plastics Engineers, Process Engineers, Quality Control Engineers, Robotic Engineers, and Traffic Engineers. Systems Engineers and Structural Engineers.

This chapter will expose you to all the types of engineering. It will define the five areas of work: research, development, application, management and maintenance of engineering. It also will define the five primary career paths that engineers follow: industry, consulting, government, academics and internet. It will enlighten you on the long history of engineers in different cultures.

The six traditional areas of engineering will be described: chemical, civil, electrical, industrial, materials science and mechanical and you will know that preparation in each of these areas will provide a solid foundation for a wide range of specialties. You will have a better idea of what area you want to go into and why.

Specialties such as aerospace, agricultural, automotive, biomedical, computer, environmental, manufacturing and petroleum will be discussed. You will know what each one entails, its history and what qualities are needed to succeed in that specialty.

CHAPTER 3: CHOOSING THE RIGHT ENGINEERING FIELD FOR YOU
- Choosing the Right Field
- How to find Out The Field You Are Most Interested In
- Your Strengths and Skills
- Code of Ethics for engineers
- Questions and answers that will help you find the right engineering field

This chapter will give you a good idea of how to find the right type of engineering job for you. It will provide you with practical advice on volunteering, summer camps, work study programs, and how to interview for engineering positions. Your personality traits and abilities will be examined to provide you with insights into what is needed in each specialty. Contractors and consultants will be defined and you will have a better idea as to where they fit into the engineering profession and if you are interested in this type of work.

Your strengths and skills are examined and evaluated to help you find the right engineering field for you. A summery of the Code of Ethics for Engineers is given in this chapter.

CHAPTER 4: EDUCATION AND LICENSING
- Picking a College
- Undergraduate and Graduate Programs
- Alternate Programs
- Engineering Licenses
- Tips for Meeting the licensing Requirements and taking the exams
- Examples of Education needed for certain Engineering fields

You will find out that you need at least a BS degree for any type of engineering and a master's degree or even a PhD for some engineering positions. How to pick the right college for you will be discussed and alternative programs evaluated. The traditional engineering majors and their specialties will be outlined and you will have a better understanding of what you should major in and why.

Licenses for engineers will be defined and you will know the different types of licenses as well as when, why and how to get them. Also you will understand when you need to be licensed and when if is not necessary. The Fundamental of Engineering, Professional Engineering and other licenses will be described. The exams for these licenses will be discussed in detail and helpful hints on passing these exams will be given.

Different types of engineering and their degree requirements and licensing will be summarized. Engineering sectors such as ceramics, environmental, computer, industrial, packaging, plastics, quality control, and traffic operations are just a few that will be described in this chapter.

CHAPTER 5: SALARIES AND POSITIONS

- Networking Rules
- Recruiters
- Alternative Position finders
- Expected Growth in traditional engineering and specialties
- Work Place and Outsourcing
- Salaries
- Information on how to find an Engineering Position
- How to find a Mechanical Engineering Job
- How to do a Job Search

This chapter will tell you how to get the position that is right for you. You will have a good idea of what type of engineering suits you and what the requirements are. It points out that the first step in finding a position is networking. Rules and etiquette involved in networking are defined. Also recruiters and how to present yourself is discussed. Alternative position finders such as summer employment, work study programs and cooperative experiences will be outlined.

The general expected growth for each type of engineering will be stated and detailed expectations will be given. Traditional engineering programs such as chemical, civil, electrical and mechanical will be reviewed. Specialties like aerospace, biomedical, computer hardware and software, environmental, industrial, materials science, mining, geological and nuclear engineering will be outlined and you will know the expected growth and employment opportunities of each.

The latest salaries for all types of engineers are given and you will have a good idea of what you would be earning at different times throughout your career. You will know what areas of the engineering profession earn the most and the least.

CHAPTER 6: MEET THE ENGINEERS

- David's Day at Work
- The Partner Personality
- Camp Made Engineer
- One of a Kind Engineer
- Love the Sport Engineer

- Opinion Engineer
- The Flying Engineer
- Seas of Opportunities
- The Proud Engineer
- Footprints in the Environment
- Senior Structural Engineer
- The Project Manager

These are some of the examples of engineers' types that will be described in this chapter. Their stories will give you a good idea of how and why they became engineers, the jobs they do and why they do it. You will learn what to expect in engineering situations and it will enlighten you and give you a realistic view of the life of different types of engineers.

The day to day schedule depicted in the opinionated engineer's story gives you a good description of what it would be like on the job as a senior manufacturing engineer.

The partner personality enables you to look at one of the original partners of a firm and what his inner feelings are.

The camp made engineer's story describes how going to engineering camp in high school influenced one person's decision to become an engineer.

The one of a kind engineer gives you insight on what it would be like to work on a first of a kind project.

These are just a few highlighted engineers that are described in this chapter. All of the engineers in this chapter give you clues to use to succeed in the engineering profession. From how to win a contract to overcoming complication to completing a project are addressed.

CHAPTER 7: TEAM WORK AND INTERDEPARTMENTAL RELATIONSHIPS
- Client Relations
- Team work
- Strategies When Starting a Position

The many relationships in the engineering profession are discussed and analyzed in this chapter. Workings with clients are described and the major points of the relationship will be outlined. Sample solutions on how to deal with resistance and other obstacles are given.

A comprehensive list of successful team strategies are outlined to help you succeed while working in a team. Self questioning procedures enable you to understand what type of team player you are and need to be.

CHAPTER 8: MEETINGS AND CONFERENCES
- Meeting and conferences
- Visual Helpers
- Guidelines for a Sales Meeting

Meetings and conferences are discussed in this chapter and you will have a better picture of what happens and what is expected of you at meetings and conferences. Group, one on one, strategy and quarterly meetings are some types that are discussed in detail so you have hints on how to prepare and shine during any type of meeting. You will discover that the more meetings you are invited to, the more valued your input is. You will learn that meetings are an important means of getting known in the company and how they help you achieve promotions or transfers.

The difference between meetings and conferences is described. You will become knowledgeable on the most common conference, the technical conference, and why it is so important. This chapter also described how conferences are usually a voluntary situation and are held in major cities. You will understand why your company would send you to a conference and what they expect from you in return. This chapter brings up the important reasons for meetings. You will learn how to use them to your advantage.

CHAPTER: 9 HOW TO MAKE YOUR COMPANY WORK FOR YOU
- How Companies Are Organized
- Companies Responsibilities
- Company Rewards
- Your Responsibilities
- Large, Midsize and Small companies
- The Right track

Every company has an organizational structure and this chapter shows it to you. Different types of companies such as small, medium and large are discussed and the different roles an engineer may have in them.

Choices of becoming a specialist or moving into management are some of the aspects described. The non existence of unions and how raises and promotions are given is outlined in this chapter. What you probably will do on you first job is described and hints on how to act for success are discussed. Steps in the advancement process are given and you will have a better idea of how advancement takes place and what to do to achieve it.

The realization of you being responsible for your own success in the company is stated. The plain fact that a corporation does not owe you very much is recognized. You will learn you must consider the goals, values, and culture of the organization that you work for as well as you own interests, values and goals. Keeping an optimistic attitude is also discussed.

Your personal and professional growth in the company and how to achieve your goals are explored. Also, the different tracks you can take such as technical verses management track are defined and you will have a better understanding of which is better for you.

CHAPTER 10: THE TEN COMMANDMENTS OF AN ENGINEER
1.   Keep Learning
2.   Pick the right college
3.   Pick the right field
4.   Never talk badly about your coworkers or your manager
5.   Make meetings work for you
6.   Make the company work for you
7.   Be a team player
8.   Get the right engineering position for you
9.   Get the right license and degree
10.   Always perform your work in a legal and ethical manner

This chapter gives details on how to fulfill The Ten Commandments of an engineer in order to achieve a successful entry into any engineering setting. Excerpts from the book will be reviewed for solutions to any problems you may encounter in the profession.

"Engineers are not boring people. They just get excited over boring things."

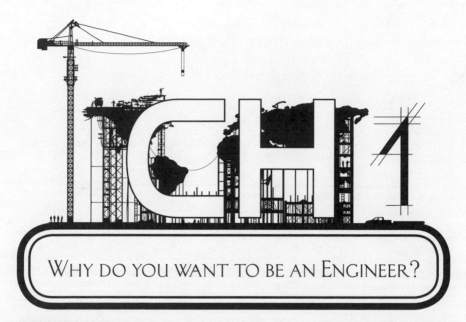

WHY DO YOU WANT TO BE AN ENGINEER?

Why do you want to be an engineer? That is the major question you should answer before you start a career in engineering. If you love math and science and have always been fascinated by "tinkering" or solving specific problems, you're probably on the right track. All engineers do not have to be brilliant but most are above average in intelligent. Creativity is important and having an analytical mind and the capacity for detail are both plusses.

The work of engineers has a more widespread impact on people than the average person thinks. Engineers have influenced discoveries and inventions that have been part of our everyday lives for generations. They use scientific knowledge and tools to design products, structures, and machines. Engineers apply theory and science to solving problems from making a better mousetrap to saving the Ozone layer, from making faster cars to devices that save lives such as heart pacemakers. Most engineers specialize in a particular area but have a knowledge base and training that can be applied to many fields. Engineers apply the theories and principals of science and mathematics to research and development of

economical solutions to technical problems. Their work is the link between perceived social needs and commercial applications.

Few careers can enable you to make money by creating things rather
then working people harder or cutting cost or people. Engineering is a
field that enables you to make something new, improve and create something while giving you a decent living.

Engineers' Personality

Are you the type of person who reacts immediately to the environment
around you? Are you observant and quick to see problems and opportunities? Are you spontaneous and prefer responding to things that are
happening right now? Do you like to take practical actions that do not
require a lot of pondering? Are you a good fixer of things or have the
ability of getting something done right away? Do you enjoy handling
emergencies or solving practical problems? Do you prefer and are you
best at activities that focus on direct observation and action?

Do you naturally approach the world in an active and exploratory way?
Are you more active then passive? Do you "seize the moment"? Are you
attracted to work in which you can learn through science and technology? If you enjoy manipulating and understanding scientific principals in
a practical way then engineering may appeal to you.

Take this quiz to learn if your personality is the personality of an engineer.

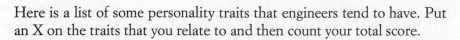

An Engineering Personality Quiz

Here is a list of some personality traits that engineers tend to have. Put an X on the traits that you relate to and then count your total score.

- ❑ 1. I prefer to manage crises
- ❑ 2. I focus intensely on a problem
- ❑ 3. I expedite an immediate solution
- ❑ 4. I'm direct and practical
- ❑ 5. I tend to leave strategic and long term planning to other people
- ❑ 6. I consider non-traditional as well as traditional approaches to problem solving
- ❑ 7. I act or lead whenever it seems expected
- ❑ 8. I push the boundaries of routine, structure, and stability
- ❑ 9. I work around as well as within the system
- ❑ 10. I have a take charge attitude
- ❑ 11. I take straight forward action
- ❑ 12. I'm independent
- ❑ 13. I don't hesitate to confront others if I disagree
- ❑ 14. I am a master at negotiations, persuasion and compromise
- ❑ 15. I remove roadblocks and keep projects moving forward
- ❑ 16. I don't get impatient with efforts to dig deeply into ongoing root causes of problems
- ❑ 17. I may focus more on expedient actions than on people
- ❑ 18. I prefer hands on activities
- ❑ 19. I enjoy competitions, challenges and taking risks
- ❑ 20. I am interested in practical applications
- ❑ 21. I get bored and will disengage when presented with abstract theory
- ❑ 22. Prefer data to opinion
- ❑ 23. I enjoy learning about and doing troubleshooting, decision making and problem solving
- ❑ 24. I don't easily become frustrated

If you checked more of these traits than not, you probable have an engineering type personality.

Know your personality

Now that you have an insight into your personality you can determine whether the engineering profession is for you. Most engineers are effective leaders when roadblocks affect their group's ability to complete tasks. Personalities that tend to ignore or avoid organizational structure and process and are nontraditional in their leadership approach may make good engineers.

If you are impatient with routine, structure and stability and are willing to work around rather than within the system if the system impedes your ability to complete a task, and prefer not to deal with the theoretical and long term or the ambiguous, engineering could be for you. If you have a commanding behavior and a take charge attitude to express your leadership and are skilled at removing, slipping under or going around roadblocks to keep projects moving forward you may have the engineer's personality.

Engineers usually are highly motivated to deal with and resolve immediate conflicts. They can be masters at negotiation, persuasion and compromise. However, if they have a difference of opinion with others they will not hesitate to confront them. They tend to analyze matters in great detail. They are logical types focusing more on expedient action rather than on considering the 'people' side of the problems.

Engineers usually prefer learning through lots of action and hands on activities. They may be experimenters or tinkerers. They most likely enjoy competition, challenge, and taking risks when learning. They are practical and want to get the straight facts about a topic. They can become bored and disengaged when presented with abstract theory unless a logical connection can be drawn between the theory and practice.

Engineering type personalities usually approach their work in a realistic manner. They are highly aware of the world around them and attend to and keep track of details in an important way. They are quick to accept the facts and see things as they are. Theory, abstraction, and metaphor are overly indirect and complex ways of seeing the world to most engineers. Theory may have some value when it can be directly applied, but preferences are initially focused on that which is concrete. They trust what they can experience directly. They value tangible products or results.

You might want to be an Engineer...

Or you might want to be an engineer if you can relate to any or most of these little excerpts on life.

So just for fun put an x on the ones you relate to.

- ❏ If you have no life – and you can prove it mathematically.
- ❏ If you chuckle whenever someone says "centrifugal force".
- ❏ If it is sunny and 70 degrees outside, and you are working on your computer.
- ❏ If you think in "math".
- ❏ If you have a pet named after a scientist.
- ❏ If you laugh at jokes about mathematicians.
- ❏ If the Humane Society has you arrested because you actually performed the Schrodinger's Cat experiment.
- ❏ If you consider any non-science course "easy".
- ❏ If the "fun" center of your brain has atrophied from lack of use.
- ❏ If you would assume a "horse" is a "sphere" in order to make the math easier.
- ❏ If you think a pocket protector is a fashion accessory.
- ❏ If you tell your spouse "I can fix that", even if you can't.
- ❏ If you go on the rides at Disneyland and sit backwards in the chair to see how they do the special effects.
- ❏ If you find yourself at the airport on your vacation studying the baggage handling equipment.
- ❏ If your IQ is bigger than your weight.
- ❏ If you remember 7 computer passwords but not your anniversary.
- ❏ If in college, you thought "Spring Break" was metal fatigue failure.
- ❏ If you are at an air show and you know how fast the skydivers are falling.
- ❏ If your favorite magazine is Popular Science.
- ❏ If you can type 70 wpm but can't read your own handwriting.
- ❏ If people groan at parties when you pick out the music.
- ❏ If you can't remember where you parked your car for the 3rd time this week.
- ❏ If you did the sound system for your senior prom.

❑ If you ever forget to get a haircut – for 6 months.

❑ If your checkbook always balances.

❑ If you got to the electronics store and the salesman asks you the questions.

❑ If your spouse says the way you dress is no reflection on them.

❑ If your wristwatch has more buttons than a telephone.

❑ If you have more friends on the internet than in real life.

❑ If you thought the real heroes of "Apollo 13" were the mission controllers.

❑ If your dress clothes come from Sears.

❑ If you think your computer looks better without the cover.

❑ If you spent more money on your PDA than on your wedding ring.

❑ If you, instead of buying groceries, end up buying software.

❑ If you say "good night" to your computer before going to bed.

❑ If you turn on your computer before the room light when you get home from work.

❑ If you know what http:// stands for

❑ If you've ever tried to repair a $5 radio.

❑ If you have a neatly sorted collection of old nuts and bolts in your garage.

❑ If you bought your spouse's Valentine Day gift at Home Depot.

❑ If your glasses are crooked on your face.

❑ If the inside of your car is covered with fast food containers, candy wrappers and apple cores.

❑ If your favorite part of the 6 o'clock news is comparing their satellite weather picture with yours.

❑ If your 3 year old son asks why the sky is blue and you actually try to explain.

❑ If you're over 30 and can program the VCR

❑ If you already spent 2 hours trying to assemble your kid's toy, but still refuse to read the instructions.

❑ If you've already calculated how much money you earn per second.

❑ If you read the Wall Street Journal for entertainment.

❑ If you check your company stock price every hour and maybe even on weekends.

❑ If you look forward to a quiet evening with our computer and a bunch of data files to analyze.

❑ If you didn't mind being called a nerd in college.

- ❑ If your family pet's name is "Laptop".
- ❑ If all you begin sentences with "What if…".
- ❑ If you mow geometric design patterns on your lawn.
- ❑ If your living room has clocks with names of different cities under them.
- ❑ If your four basic food groups are caffeine, fat, sugar, and chocolate.
- ❑ If you can't follow a recipe without a chemistry scale.
- ❑ If your spouse sends you an email instead of calling you to dinner.
- ❑ If you look forward to Christmas only to put your kids toys together.
- ❑ If you window shop at Radio Shack.
- ❑ If you have ever taken the back off you TV just to see what's inside.
- ❑ If you have ever burned down the school gym with your Science Fair project.
- ❑ If you are currently gathering the components to build your own nuclear reactor.
- ❑ If you have ever saved the power cord from a broken appliance..
- ❑ If you own a set of itty-bitty screw drivers, but you don't remember where they are.
- ❑ If you have more toys than your kids.
- ❑ If you need a checklist to turn on the TV.
- ❑ If you make a copy of this list and post it on your door.
- ❑ I hope you had fun analyzing yourself and maybe even gotten some insight into your personality.

The Career Choice Process

Deciding if the engineering profession is for you involves a career choice process through which you will make some major decisions about you life. There are two processes involved. The first step in the career choice process is self analysis, determining your personal goals. The second process is the search, which includes your natural professional abilities to become an engineer.

You should approach the career choice process in a rational way and take certain steps in sequence. If you are not sure about which way your professional life should go you must organize your thoughts. It is important that you have enough self awareness and self knowledge to be able to make a decision that will best satisfy your personal goals as well as your abilities, skills, needs, and values.

Not all engineers need to master the same set of professional skills. This depends on what type of engineering you go into. In general, engineers apply the theories and principals of science and mathematics to research and development of economical solutions to technical problems along with general problem solving abilities and creativity. As a future engineer you should feel comfortable with technology such as digital prototyping software and computer aided design. The work is a link between perceived social needs and commercial applications.

Career theory suggests that people enjoy using the skills they have and that they will do well in the future in activities utilizing these skills. Take a piece of paper and on one side write down what your personal abilities, skills, needs, values, and goals are. Then on the other side of the paper write down what abilities, skills, needs, values, and goals you think an engineer will need. This kind of a chart is called a 'T-Chart' and was popularized by Ben Franklin, one of our nation's first engineers.

Do the sides match up? What do you need to work on to be an engineer? Do you have most of the qualities? Would you be working on more of the skills then you think you should be? You want to achieve a balanced happy life; will becoming an engineer achieve that for you?

What Do You Expect From the Engineering Profession?

What do you expect from the engineering profession? Answer the following questions and note your answers. These questions will help you evaluate what you expect from the engineering profession.

1. How large a part of wanting to be an engineer is your desire to work within the fields of science and mathematics?

2. How important is salary and job security to your desire to be an engineer?

3. Are you willing to commit a lot of time, money, and effort to education and training?

4. What kind of work environment do you want to work in?

5. How much stress can you handle on the job?

6. Do you want to be your own boss, run a business with others, or will you be content to be a salaried employee?

7. Do you have a desire to create things?

8. Are you a detailed person and a logical thinker?

9. Can you work as part of a team and communicate well with others?

10. Do you keep careful records?

By completing this exercise and writing your answers down, you now have more clear and focused career expectations. Below is a review of each of the questions posed above and a discussion of how each issue comes to play in the engineer field.

(1) The engineering profession is a science and mathematics industry. If you love math and science and have always been fascinated by solving specific problems then engineering is probably for you.

(2) You are not going to become rich being an engineer, but engineering majors command one of the highest salaries offered among new graduates and the starting salaries inch up every year. Average starting salaries range from $35,000 to $75,000. Often an experienced engineer makes only a few thousand dollars more than a recent graduate and salary growth tends to remain sluggish after that. If earning a lot of money is important to you, remember that the largest number of CEO's of the Fortune 500 companies have Engineering degrees and even if managing others is not your forte, earnings are unlimited in the area of technical sales. Now-a-days, no job is secure. Engineers working for government agencies might have more security than engineers working in private industry but they also earn less. If you are good at what you do, you won't have any problems. Even if you find yourself sometime in the future being laid off, you'll note that there are more Engineering listings in the professional classified sections of the major newspapers and there are more recruiting firms specializing in engineering than any other types of professions. In general engineers are mobile and willing to relocate for career opportunities.

(3) You will you need a bachelor degree for all engineering fields. It is also important for engineers to continue their education throughout their careers because much of their value to employers depends on knowledge of the latest technology. Most companies offer paid training for their engineering staff, and virtually all offer reimbursement for taking college courses within your field.

(4) A career in engineering provides you with many opportunities for work settings. Most engineers work in office buildings, laboratories, or industrial plants. Others may spend time outdoors at construction and

production sites where they monitor or direct operations or solve onsite problems. Some engineers travel extensively to plants or worksites. There is a good chance you will be able to find a position in any setting that suits your needs.

(5) Any position can be stressful but depending on what type of engineering you go into the stress level can be more then the average job. At times deadlines or design standards may bring extra pressure to a job, sometimes requiring engineers to work long stressful hours. Engineering tends to be project oriented meaning that your work on a particular project must be completed by a certain deadline regardless of the hours you need to put in to get it done.

(6) Most engineers are employees working for large companies. Some engineers become private consultants. This is usually a self employed career. Some Engineers ultimately become managers while others may have opportunities in marketing or sales.

(7) No other profession is as involved in creating things, from the smallest electronic devices to the largest buildings in the world. You don't have to be creative to be an engineer but it would be helpful if you were. Having an analytical mind and the capacity for detail are both plusses. An engineering career can bring you creative stimulation. You will face new issues everyday, meet new people with new problems. It will not be boring because you will have the flexibility to change what you are doing and use your training. Engineers are problem solvers who rely on the discoveries of mathematics and science to improve the world in which we all live.

(8) You need a detailed and logical mind to become an engineer. Engineers tend to analyze problems in great detail. They are logical thinkers focusing more on expedient action rather than on considering the 'people' side of the problems they face. Engineering type personalities usually approach their work in a realistic manner. They are highly aware of the world around them and attend to and keep track of details in an important way. They are quick to accept the facts and see things as they are.

(9) You may want to look into working for a small or midsized company. There are a number of opportunities for new engineers in these companies. If you have an entrepreneurial spirit and want to find exciting opportunities this might be for you. When managed correctly these size companies can become the conglomerates of the future. However it is still important to research them well because if they do not make it you could be out of a job faster and with less security than if you were with a large organization. Here are some suggestions. Research the owner's background and reputation through local professional organizations. Research the technology and/or the process. Does the owner hold the patent for the technology you will be using? Are there any financial records available for review? It pays to be cautious but the challenge and excitement of being part of a small, growing company is something that should be seriously considered. If you are interested in large companies many corporations hold job fairs at colleges around the country.

(10) If you find yourself working on products that others will use, it is important to leave no stone unturned. Some people have a tendency to complete 95% of a project and then abandon it in favor of something new or exciting. This will not work in engineering. You must be data driven.

Get Company Information

It is wise to get as much information as you can about the company that you are interested in. If they are a public corporation, their annual report is an extremely helpful piece of information.

An engineering career can bring you satisfaction, intellectual and creative stimulation and a reasonable income. You will face new issues everyday, meet new people with new problems. It will not be boring because you will have the flexibility to change what you are doing and use your training. So evaluate these questions and answers and keep them in mind. They will help you decide if engineering is for you.

Definition of an Engineer

The American Society of Engineering Education says, "Engineering is the profession in which knowledge of the mathematical and natural sciences gained by study, experience and practice, is applied with judgment to develop ways to utilize the materials and forces of nature, economically for the benefit of mankind.

Engineers design products and machinery to build those products, plants in which those products are made, and the systems that ensure the quality of the products and the efficiency of the workforce in the manufacturing process. Engineers design, plan, and supervise the construction of buildings, highways, and transit systems. They develop and implement improved ways to extract, process, and use raw materials such as petroleum and natural gas. They develop new materials that both improve the performance of products and take advantage of advances in technology. They harness the power of the sun, the earth, atoms, and electricity and magnetism for use in supplying the nation's power needs, and create millions of products using power. They analyze the impact of the products they develop or the systems they design as well as the environment and people using them.

In addition to design and development, many engineers work in testing, production, or maintenance of products designed by other engineers. These engineers supervise production in factories, determine the root causes of breakdowns, and test manufactured products to maintain their quality. They also estimate the time and cost to complete projects. Some move into engineering management, general management, law or into sales. In sales, an engineering background enables them to discuss technical aspects and assist in product planning, installation, and use.

Engineering knowledge is applied to improving many things, including the quality of healthcare, the safety of food, and the operation of financial systems. Most engineers specialize. An engineering background provides a good base to change careers and career focus. An Environmental engineer may work in new construction or in project planning or as an advisor to the legal profession in crafting regulations and laws. More than twenty five major specialties are recognized by professional engineering societies, and the major branches have numerous subdivisions. Some examples include structural and transportation engineering, which are subdivisions of civil engineering and ceramic, metallurgical and polymer engineering, which are subdivisions of materials science engineering.

Engineers also may specialize in one industry, such as motor vehicles, or in one field of technology, such as turbines or semiconductor materials. Some other branches of engineering are, aerospace, agricultural, bio-medical, chemical, civil, computer software, electrical and electronics, environmental, industrial, mechanical, mining and geological, nuclear and petroleum, and marine engineering.

Engineers may advance to become technical specials or generalists who supervise a staff or team of engineers and technicians. Some may eventually become engineering managers or enter other managerial or sales jobs.

Engineers are creative problem solvers who rely on the discoveries of mathematics and science to improve the world in which we live.

Conclusion

After reading this chapter you are more aware of why you want to be an engineer and the many reasons for choosing the engineering profession. If you love mathematics and science and have always been fascinated with tinkering, you now know you are heading toward the right profession.

You have examined your personality and you now have the insight to determine whether the engineering profession is for you. You have reviewed the two career choice processes that are involved, self analysis and personal goals and searched your natural professional abilities to determine your potential for becoming an engineer.

We have delved into what you should expect from the engineering profession and self exploratory questions helped you to have a clear and focused career expectation.

The definition of an engineer, from the American Society of Engineering Education, gave you a broad idea of the engineering profession. I hope the answers to these questions lead you to the engineering profession.

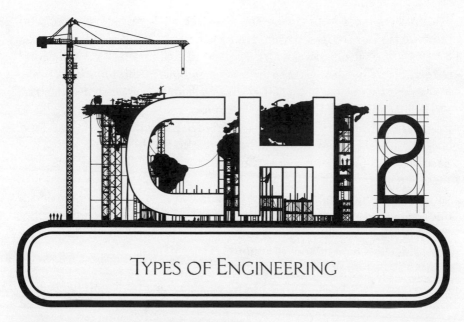

TYPES OF ENGINEERING

Five Areas of Work

All engineers engage in one or more of six areas of work: research, development, application, production, management, and maintenance. Engineers who work in research are responsible for investigating new materials, processes, or principles for practical applications of ideas and materials. Engineers who work in development use the research results to determine how best to apply them to their practical functions. Application engineers produce the actual materials, machines, and methods designed by research and development engineers. Manufacturing and maintenance engineers keep the developed idea working and make improvements and adjustments. Production engineers perform technical work related to manufacturing, production, documentation and other facets of the manufacturing environment.

Engineering is a career that has emerged as a result of many inventions of the twentieth century that dramatically changed the way people live.

While engineering has changed and grown in recent years, engineering is a very established career field with a long and distinguished history.

The first engineer known by name and achievement is Imhotep, who built the famous stepped pyramid in Egypt circa 2550 B.C. The Persians, Greeks, and Romans, along with the Egyptian civilization took engineering to remarkable heights by using arithmetic, geometry and physical science. Many famous ancient structures that are still standing today demonstrated the ingenuity and skill of these early pioneers of engineering Medieval European, like their more ancient counterparts, combined military and civil skill to carry construction to heights unknown by the ancients. They developed techniques known as the Gothic arch and flying buttress.

In Japan, China, India and other Far Eastern areas, engineering developed separately but similarly. Sophisticated techniques of construction, hydraulics and metallurgy practiced in the Far East led to the impressive beautiful cities of the Mongol Empire.

In 1747, the first use of the term civil engineer coincided with the founding in France of the first engineering school, the National School of Bridges and Highways.

Engineering Career Paths

There are five primary career paths that engineers follow: industry, consulting, government, academic and internet career paths.

Most industrial settings are high tech workplaces demanding high levels of engineering expertise to solve problems related to researching, developing and designing new products and then to manufacturing those products in a cost efficient way. Some of the other areas of industry in which engineers tend to work are accounting and finance, administration, information systems, marketing and sales and technical professional services. Many engineers in industry have obtained advanced degrees in business administration during the course of their employment. The industrial career path can lead engineers to the executive level in many companies.

Engineers who work for engineering consulting firms perform engineering tasks for other companies or organizations and when their job is done they move on to a new project with another company or organization. They work on numerous projects with different types of organizations and people. Some engineers pursue consulting careers early to help them decide where they ultimately want to work, while others prefer the changes and challenges of constantly changing projects and organizations.

The government career path can be an interesting choice. The federal, state, and local governments can be your employers. They employ many types of engineers in the space program because of the great diversity and creativity of its projects. The US Environmental Protection Agency and the U.S. Army Corps of Engineers employ many civil and environmental engineers; there are also numerous career paths for these engineers in state departments of transportation and environmental protection agencies throughout the country.

The Food and Drug Administration is a major employer of biomedical engineers while the U.S. Department of Defense continues to employ a wide variety of engineers in both civilian and enlisted positions for its agencies and installations throughout the world.

Most engineers who pursue the academic path have received either a master's or PhD. degrees and teach in colleges and universities. Some engineers decided to obtain state teaching certificates after completing their bachelor's degree and teach mathematics or science in middle schools or high schools.

The internet career path has opened new possibilities for engineers. There are opportunities for engineers with companies that are associated with the internet. There are also Internet career possibilities with more traditional companies. Companies have realized that to be successful they need to use the Internet to get their products to their customers more quickly and more cost effectively.

Traditional Areas of the Profession

There are six areas of engineering that form the traditional areas of the profession. These areas are, chemical, civil, electrical, industrial, materials science and mechanical engineering. Preparation in any one of these areas will provide a solid foundation for a wide range of specialties.

Over the years, each traditional branch of engineering has developed into focused specialties. Today, some of these specialties have become engineering professions in their own right such as aerospace, agricultural, automotive, biomedical, computer, environmental, manufacturing, mining, traffic, metallurgical, nuclear, optics, petroleum, plastics, packaging, quality control, and robotics.

Here are definitions of some major engineering sectors. This will give you an idea of what engineering you might be interested in. Engineers specialize; you will not be able to do every type of engineering. However if you are not sure of which branch of engineering you are really interested in you have the luxury of time. Most engineering curricula follow a common set of courses for the first three to four semesters.

Aerospace Engineers

Aerospace engineering encompasses the fields of aeronautical (aircraft) and astronautically (spacecraft) engineering. Aerospace engineers work in teams to design, build, and test machines that fly within the earth's atmosphere and beyond. Although aerospace science is a very specialized discipline, it is also considered one of the most diverse and hardest fields. This field of engineering draws from such subjects as physics, mathematics, earth science, aerodynamics, materials science, electronics, and biology. Some aerospace engineers specialize in designing one complete machine, perhaps a commercial aircraft, whereas others focus on separate components such as for missile guidance systems.

The creation of aircraft and spacecraft involves many branches of engineering but aerospace engineers are responsible for the total design of the craft, including its shape, performance envelope, and propulsion and guidance control systems. Professional responsibilities vary widely depending on the specific job description. The dedicated and brilliant students usually become these types of engineers.

The education and training involves how materials and structures perform under tremendous stress. In general, they are relied upon to apply their knowledge of propulsion, aerodynamics, thermodynamics, fluid mechanics, flight mechanics and structural analysis. Less technically scientific issues such as cost analysis, reliability studies, maintainability, operations research, marketing and management are also used.

There are other professional titles given to aerospace engineers such as aeronautical, astronautically, analytical, stress analysts, computational fluid dynamics (CFD), design aerospace, manufacturing aerospace, materials aerospace and marketing and sales aerospace engineers.

Aeronautical engineers work with aircraft systems and astronautical engineers specialize in spacecraft systems. Analytical engineers use engineering and mathematical theory to solve questions that arise during the design phase. Stress analysts determine how the weight and loads of structures behave under a variety of conditions.

The analysis is performed with computers using complex algorithms. Fluid dynamics engineers use sophisticated high speed computers to develop models used in the study of fluid dynamics. Design aerospace engineers draw from the expertise of many other specialists while devising the overall structure of components and entire crafts, meeting the specifications developed by those more specialized in aerodynamics, astrodynamics and structural engineering.

Materials aerospace engineers determine the suitability of the various materials that are used to produce aerospace vehicles. Marketing and sales aerospace engineers work with customers, usually industrial corporations and the government, informing them of product performance. They act as a liaison between the design engineers and the clients to help ensure that the products delivered are performing as planned. Sales engineers also need to anticipate the needs of the customer to inform their companies of potential new marketing opportunities.

History of Aerospace Engineering

The roots of aerospace engineering can be traced as far back to when people first dreamed of flying. Thousands of years ago, the Chinese flew kites and later experimented with gunpowder as a source of propulsion. In the 15th century, Renaissance artist Leonardo da Vinci created drawings of two devices that were designed to fly. One, the ornithophter, was supposed to fly the way birds do, by flapping its wings, the other was designed as a rotating screw, closer in form to today's helicopter.

In 1783, Joseph and Jacques Montgolfier of France designed the first hot air balloon that could be used for manned flight. In 1799, an English Baron, sir George Cayley, designed an aircraft that was one of the first not to be considered, "lighter than air", as balloons were. He developed a fixed wing structure that led to his creation of the first glider in 1849. Much experimentation was performed on gliders and the science of aerodynamics through the late 1800s.

By 1903, the first mechanically powered and controlled flight was achieved using a craft designed by Orville and Wilbur Wright. Airplane development began in earnest during World War I. In the early years of the war, aeronautical engineering encompassed a variety of engineering skills applied directly toward the development of flying machines. Civil engineering principles were used in structural design while early airplane engines were designed by automotive engineers. Aerodynamic design itself was primarily empirical with many answers coming from liquid flow concepts established in marine engineering.

The evolution of the airplane continued during both world wars, with steady technological developments in materials science, propulsion, avionics, stability, and control. Airplanes became larger and faster. Robert Goddard developed and flew the first liquid propelled rocket in 1926. The technology behind liquid rocket propulsion continued to evolve and the first U.S. liquid rocket engine was tested in 1938. More sophisticated rockets were eventually created to enable aircraft to be launched into space. The world's first artificial satellite, Sputnik 1, was launched by the Soviets in 1957. In 1961, President John F. Kennedy urged the United States to be the first country to put a man on the moon; on July20, 1969, astronauts Neil Armstrong and Edwin 'Buzz' Aldrin Jr. accomplished that goal.

Aerospace engineers should enjoy completing detailed work, problem solving, and participation in team efforts. Mathematical, science and computer skills are a must. You have to be able to communicate your ideas, share in teamwork and visualize the form and function of structures. Curiosity, inventiveness and the willingness to continue learning from experience as well as paying close attention to details are good qualities to have.

Biomedical Engineers

Biomedical engineers are highly trained scientists who use engineering and life science principles to research biological aspects of animal and human life. They develop new theories and they modify, test, and prove existing theories on life systems .They design health care instruments and devices, design prosthetic devices or apply engineering principles to the study of human systems.

Using engineering principles to solve medical and health related problems, the biomedical engineer works closely with life scientists, members of the medical profession, biologists and chemists. Most of the work areas are in research, design and teaching.

Biomedical research is multifaceted. It calls for applied knowledge of mechanical, chemical and electrical engineering as well as anatomy and physiology in the study of living systems. Using computers, biomedical engineers apply their knowledge to develop mathematical models that simulate physiological systems.

In biomedical engineering, design of medical instruments and devices are developed. Engineers work on artificial organs, ultrasonic, magnetic resonance and other kinds of imagery, cardiac pacemakers and surgical lasers. They design systems that will update hospital, laboratory and clinical needs. They also train healthcare personnel in the proper use of this new technology. The teaching aspect of biomedical engineering is on the university level.

Within biomedical engineering you can concentrate in different areas such as bioinstrumentation, biomechanics, biomaterials, systems physiology, clinical engineering and rehabilitation engineering. Biomechanics is mechanics applied to biological or medical problems such as artificial hearts, kidneys and joints. System physiology uses strategies to gain a comprehensive understanding of living organisms ranging from simple bacteria to humans. Rehabilitation engineering's goal is to improve the quality of life for individuals with physical impairments. They often work with the disabled and modify equipment for their individual use.

History of Biomedical Engineering

In its broadest sense, biomedical engineering has been with us for centuries, perhaps even thousands of years. In 2000, German archeologists uncovered a 3,000-year-old mummy from Thebes with a wooden prosthetic tied to its foot to serve as a big toe. Even more recently, skulls discovered in Pakistan and dated to be between 7,500 and 9,000 year old were found to have precision holes drilled in some of the teeth and is believed to be an example of the earliest form of dentistry. Egyptians also used hollow reeds to look and listen to the internal goings on of the human anatomy. In 1816, modesty prevented French physician Rene Laennec from placing his ear next to a young woman's bare chest, so he rolled up a newspaper and listened through it, triggering the idea for his invention that led to today's ubiquitous stethoscope.

No matter what the date, biomedical engineering has provided advances in medical technology to improve human health. Biomedical engineering achievements range from early devices, such as crutches, platform shoes, wooden teeth, and the ever-changing cache of instruments in a doctor's black bag, to more modern marvels, including pacemakers, the heart-lung machine, dialysis machines, diagnostic equipment, imaging technologies of every kind, and artificial organs, implants, and advanced prosthetics.

As an academic endeavor, the roots of biomedical engineering reach back to early developments in electrophysiology, which originated about 200 years ago. An early landmark in electrophysiology occurred in 1848 by DuBois Reymond .Raymond's contemporary, Hermann von Helmholtz, is credited with applying engineering principles to a problem in physiology and identifying the resistance of muscle and nervous tissues to direct electrical current.

Between World War I and World War II, a number of laboratories undertook research in biophysics and biomedical engineering. Only one offered formal training: the Oswalt Institute for Physics in Medicine, established in 1921 in Frankfurt, Germany, a forerunner of the Max Planck Institute fur Biophysik.

Following the Second World War, administrative committees began forming around the combined areas of engineering, medicine and biology. A biophysical society was formed in Germany in 1943.

In 1968, the Biomedical Engineering Society was formed to give "equal status to representatives of both biomedical and engineering interests and promote the increase of biomedical engineering knowledge and its utilization. The American Institute for Medical and Biological Engineering was created in 1992. The earliest academic programs began to take shape in the 1950s. Ultimately these three universities, the Johns Hopkins University, the University of Pennsylvania and the University of Rochester were among the first to win important training grants for biomedical engineering from the National Institutes of Health.

During the late 1960s and early 1970s, development at other institutions followed similar paths, but occurred more rapidly in most cases due to the growing opportunities in the field and in response to the important NIH initiative to support the development of the field. The earlier institutions were soon followed by a second generation of biomedical engineering programs and departments. These included: Boston University in 1966; Case Western Reserve University in 1968; Northwestern University in 1969; Carnegie Mellon, Duke University, Rensselaer and a joint program between Harvard and MIT in 1970; Ohio State University and University of Texas, Austin, in 1971; Louisiana Tech, Texas A&M and the Milwaukee School of Engineering in 1972; and the University of Illinois, Chicago in 1973.

A major development took place in late 2000 when President Bill Clinton signed a bill creating the National Institute of Biomedical Imaging and Bioengineering at the NIH.

You should have a strong commitment to learning if you plan on becoming a biomedical engineer. You should be scientifically inclined and be able to apply that knowledge to problem solving. Becoming a biomedical engineer requires long years of schooling because you need to be an expert in engineering and biology. Also you have to be familiar with chemical, materials science, and electrical engineering as well as physiology and computers. Biomedical Engineers tend to team with medical doctors.

Ceramics Engineers

Ceramics engineers work with non-metallic materials such as clay and inorganic elements such as zirconium and silicon. They are part of the ceramics and glass industry which manufactures such common items as tableware as well as highly technical items such as ceramic tiles for the space shuttle and superconducting materials. Ceramic engineers perform research, design machinery and processing methods as well as develop new ceramic materials and products.

The ceramics engineer is a materials science engineer who works toward the development of new products. They also use their scientific knowledge to anticipate new applications for existing products.

Ceramics research engineers conduct experiments and study the chemical and physical properties of materials as they develop the ideal mix of elements for each new product. Many research engineers are fascinated by the chemical, optical, physical and thermal properties and interactions of the oxides that make up many ceramic materials.

Ceramics design engineers take the information accumulated by researchers and develop actual products as diverse as Corel dinnerware to optical fiber to high temperature engine parts. In addition to working on new products they may design new equipment to produce the products.

Ceramics test engineers test materials that have been chosen by the researchers to be used as sample products. Other ceramics engineers might be involved in hands-on work, such as grinding raw materials and firing products. Maintaining proper color, surface finish, texture, strength and uniformity are some responsibilities of the ceramics engineer.

History of Ceramics engineering

Thousands of years ago ceramics makers relied on one raw material, clay. Originally clay was dried in the sun to harden before use. By 5,000 BC it was being fired to make it more durable but not much further advancement was made in its development for thousands of years.

The scientific and industrial revolution of the 19th century enabled people to begin using ceramics in complex scientific and industrial processes. New manmade materials were also developed which made possible the development of new products that were stronger, more or less transparent, less brittle, or more magnetic. Early twentieth century ceramics engineers used porcelains for high voltage electrical insulation.

Today basic ceramic materials such as clay and sand are being used not only by craftspeople but also to create memory storage, optical communications, and electronics. Ceramics engineers are now working with advanced materials such as silicon carbides, nitrides, and fracture resistant zirconia's.

If you are interested in the materials we use every day and how to improve them, then ceramic engineering may be for you. You will have to have knowledge of material composition and how to mix different mediums to products quality materials. Ceramics Engineering is a specialty within the general field of materials science.

Chemical Engineers

Chemical engineers are involved in evaluating methods and equipment for the mass production of chemicals and other materials requiring chemical processing. They may also be involved in developing new chemical compounds from scratch. They also develop products from materials such as plastics, metals, petroleum, detergents, pharmaceuticals, and foods. They improve environmentally sound processes, determine the least costly production method, and formulate the material for easy and safe use.

Chemical engineering is one of four major disciplines in addition to electrical, mechanical, and civil engineering. Chemical engineers are trained in many fields in addition to chemistry such as physics, mathematics and other sciences such as biology and geology. They are one of the most versatile of engineers with many specialties and employed in many different industries.

Research engineers work with chemists to develop new processes and products. Others run tests on the processes and make any necessary adjustments to achieve the consistency and form they are looking for. They strive to improve the processes, reduce hazards and waste, while cutting production time and cost.

Process design engineers determine how the product can most efficiently be produced on a large scale while still guaranteeing high quality results. Besides working on the steps of the process, they often assist in the plant and equipment design along with mechanical, electrical and civil engineers.

Project engineers oversee the construction of new plants and the installation of new equipment. In construction, chemical engineers may work as field engineers who are involved in testing and initial operation of the equipment. Chemical engineers working in environmental control areas are involved in waste management, recycling, and control of air and water pollution. Technical sales engineers may work with customers of manufactured products to determine what products can be used more economically or why a product is not working properly.

History of Chemical Engineering

Chemical engineering existed in early civilization as applied chemistry. Ancient Greeks distilled alcoholic beverages as did the Chinese who by 800 BC had learned to distill alcohol from the fermentation of rice. Aristotle the fourth century Greek philosopher wrote about a process for obtaining fresh water by evaporating and condensing water from the sea. The foundations of modern chemical engineering started during the Renaissance when experimentation and questioning of accepted scientific theories became widespread. Many new chemical processes were developed such as producing sulfuric acid for fertilizers and textile treatment, and using alkalis for soap. Alchemy was also a form of chemical engineering.

The atomic theories of John Dalton and Amedeo Avogadro, developed in the 1800's supplied the theoretical underpinning for modern chemistry and chemical engineering.

With the inventions of large scale manufacturing in the mid 19th century modern chemical engineering began to develop. Chemical manufacturers soon required chemists knowledgeable in manufacturing processes. These early chemical engineers were called chemical technicians or industrial chemists. The first course in chemical engineering was taught in 1888 at the Massachusetts Institute of Technology and by 1900 chemical engineers had become a widely used job title.

You should be knowledgeable in the four major disciplines, electrical, mechanical and civil in order to be a chemical engineer.

Civil Engineers

Civil engineers are involved in the design and construction of the physical structures that make up our infrastructure; roads, bridges and tunnels, buildings, sewers and water distribution, and airstrips and harbors. It involves knowledge applied to the practical planning of the layout of our cities, towns, and other communities. It is concerned with modifying the natural environment to better the lifestyles of the general public. A specialized branch of Civil Engineering is also as structural engineering. Civil engineers use their knowledge of materials science and theory, economics and demographics to devise, construct and maintain our physical surroundings. They apply their skills of other branches of science such as hydraulics, geology, and physics to their work.

Surveying and mapping engineers determine the best sites and approaches for construction. These engineers use satellite imagery and electronic instruments like GPS systems and ground penetrating radar to measure the area and conduct underground probes for bedrock and ground water.

Many civil engineers work strictly as consultants on projects usually specializing in one area of the industry such as water systems, transportation systems, or housing structures. Clients can be individuals, corporations or government agencies. Civil engineers will often work the architects and translate designs into functional structures.
Other civil engineers work mainly as contractors or consultants and are responsible for the actual building of the structure. They are known as construction engineers. During construction, these civil engineers are

the lead project manager and must supervise the labor and make sure the work is completed correctly and efficiently according to government specifications and standards. After the project is completed, they must set up maintenance schedules and check the structure for a certain length of time. Later, the maintenance and repair is often given to local engineers. Civil engineers may be known by their area specialization for example as transportation engineers.

Civil engineering is a nice stable job for the good at math, responsible, intelligent personality who really wants to find a decent paying where ever he desires to live.

History of Civil Engineering

The origins of civil engineering date back to ancient Greek, Egyptian. Structures there offer the first known examples of a plan to build, and the use of mathematics to achieve a desired result instead of building haphazardly. Early Civil Engineers were most often employed for military action such as building bridges across rivers or building and maintenance of "Siege Machines" for the scaling of fortifications and walls. The evolution of civil engineering can also be traced back to Roman and medieval times. Several bridges made of timber were built over the Thames River. The Romans also created a series of aqueducts on a scale never seen before to bring water into the city of Rome. But the first person on record to call himself a civil engineer was an eighteenth century Englishman, John Smeaton (1724-1792), who sought to distinguish civilian construction work from that of military engineers.

The oldest of all engineering disciples is Civil Engineering. In America, Civil Engineering could be deemed to have begun in 1776 when Congress authorized an "Engineer Corps". In 1821, the first reference to a developing distinction between "military" and "civilian" engineering activities occurred when Congress directed certain surveys of roads and canals be made by "Engineer Officers" and "Civil Engineers". West Point first offered a course entitled "Civil Engineering" in 1823. In 1835, the first Engineering degree was conferred by the first technological degree-granting university in the English speaking world, Rensselaer Polytechnic Institute (New York).

As the world's various civilizations have progressed through the ages, civil engineering expertise has developed significantly and has shaped society in profound ways — from the pyramids of Egypt to the Petronas Towers in Malaysia.

As the population of the world continues to grow and communities become more complex, civil engineers must remodel and repair structures. New highways, buildings, airstrips, etc. must be designed to accommodate ever changing public needs. Today, more and more civil engineers are involved with water treatment plants, and the natural environment as is evident in the growing number of engineers working on preserving wetlands, maintenance of national forests and restoration of sites around mines, oil wells and industrial factories. Much of this work is done by the Army Corps of Engineers.

Civil engineers have one of the world's most important jobs. They build our quality of life. With creativity and technical skill, civil engineers plan, design, construct and operate the facilities essential to modern life, ranging from bridges and highway systems to water treatment plants and energy-efficient buildings.

Civil engineers are problem solvers, meeting the challenges of pollution, traffic congestion, drinking water and energy needs, urban redevelopment and community planning. As the technological revolution expands, as the world's population increases, and as environmental concerns mount, civil engineering skills will be increasingly needed throughout the world. It can truly be said that we cannot have modern civilization without civil engineers.

Electrical & Electronics Engineers

Electrical engineers apply their knowledge of the sciences to work with equipment that produces and distributes electricity such as generators, transmission lines and transformers. They also design, develop, and manufacture electric motors, electrical machinery and ignition systems for automobiles, aircraft and other types of engines. In the late 1950's, the field was broken in two creating the Electronics Engineer and continuing the Electrical Engineer. Electronics engineers are concerned with devices made up of electronic components such as computers, telephones and radios. Both types of engineers are often referred to as electrical engineers.

Electrical and electronics engineering is a diverse field with numerous divisions and departments within which engineers work. In general they use their knowledge of science in the practical applications of electrical energy. They can work on atom smashing or small microchips as well as large power generation equipment as is needed at atomic energy or other electrical generation plants. They are involved in invention, design, construction and operation of electrical and electronic systems and devices of all kinds.

There are many specialties such as design and testing, research and development, production, field service, sales and marketing, and teaching. Even within each specialty there are divisions. Researchers develop potential applications; they conduct tests and perform studies to evaluate fundamental problems. Those working with design and development adapt the researchers' findings to actual practical application. Production engineers may have the most hands on tasks in the field. They take care of materials and machinery, schedule technicians and assembly workers and make sure that standards are met and products are quality controlled.

Field service engineers act as liaisons between the manufacturers, distributors and the clients after the systems are in place. A sales engineer contacts clients and learns about the clients' needs and reports back to the various engineering teams. Whatever type of project you work on there will be a fair amount of desk work like writing status reports and simulating designs on computers. This will vary from project to project. Electrical and electronics engineers' work touches almost every aspect of our lives, the lights in a room, cars on the road, televisions, stereo systems, telephones, and computers are just a few. They have had a say in science, industry, commerce, entertainment and even the arts.

History of Electrical and Electronics Engineering

Electrical and electronics engineering had their beginnings in the 19th century. In 1800, Alexander Volta (1745-1827) made a discovery that opened a door to the science of electricity. He found that electric current could be harnessed and made to flow and do useful work.

Electricity was also discovered by Benjamin Franklyn. By the mid 1800s the basic rules of electricity were established and the first practical application appeared. At that time Michael Faraday (1791-1867) discovered the phenomenon of electromagnetic induction. Further discoveries followed. In 1837 Samuel Morse (1791-1872) invented the telegraph and in 1876 Alexander Graham Bell (1847-1922) invented the telephone. The incandescent lamp was perfected by Thomas Edison (1847-1931). In 1878 the first electric motor was invented by Nicholas Tesla (1856-1943) although Faraday had built a primitive model of one in 1821. It is worth noting that Western Union, the first company to commercially exploit the telegraph, closed it's doors early in 2006 because of the email and internet revolution. Actually Western Union still exists but not as a communication company but for transferring money.

These inventions led to dependence on electricity. In the 19th century the start of electronics began, which is different from the science of electricity by its focus on lower power generation. In the late 1800s, current moving through space was observed for the first time which was called the "Edison effect". This lead to the potential transmission of electromagnetic waves for communication. The first practical demonstration of radio was made by Marconi. The original transmission site can still be seen on Cape Cod, Massachusetts.

Prior to the Second World War, this type of engineering was commonly known as radio engineering and basically was restricted to aspects of communications and radar, commercial radio and early television. Later, in post war years, as consumer devices began to be developed, the field grew to include modern television, audio systems, computers and microprocessors. In the mid to late 1950s, the term radio engineering gradually gave way to the name electronics engineering. In the 1950s, transistors were built on tiny bits of germanium and later silicon. Several transistors were integrated onto a single piece of silicon to become what we call integrated circuits or microchips. The computer industry is a major beneficiary of the creation of these integrated circuits.

Obsolescence of technical skills is a serious concern for electrical engineers. Membership and participation in technical societies, regular reviews of periodicals in the field and a habit of continued learning are therefore essential to maintaining proficiency. Professional bodies of note for electrical engineers include the Institute of Electrical and Electronics Engineers (IEEE).

Hardware Engineers

Hardware engineers, also called physical designers, design, build and test computer hardware such as computer chips, circuit boards and computer systems. They also design peripheral devices such as printers, scanners, modems, and monitors. Many of these engineers are employed by major computer companies like Hewlett Packard, Dell, and Apple Computer. Hardware engineers also design radios, televisions, telephone systems, electronic control systems and sensors or any other system that requires the interconnection of integrated circuits to perform specific functions. Today's new Plasma or liquid crystal televisions require electronics engineers to work side by side with chemical and materials science engineers.

Computer engineering is broken into two major groups. Hardware engineers mainly concern themselves with the design of the actual chips and circuit boards that makeup a computer or other instrument while, software engineers design the instructions or programs as they are known that ultimately make them useful in our lives.

Electronics Engineers work with the physical parts of computers, such as computer processing units, motherboards, chip sets, video chip sets and cards, cooling units, magnetic tape, disk drives, storage devices, wired and wireless network cards as well as all the components that connect them. At the same time, mechanical engineers design the cases that the electronics will ultimately go into. Hardware engineers design and create prototypes for testing using schematic diagrams usually drawn on computers. Technicians usually assemble the parts while the engineers re-work and develop the parts through multiple testing procedures. Once the final design is completed hardware engineers oversee the manufac-ture and installation of parts.

Computer hardware engineers also work on keyboards, printers, moni-tors, mice, track balls, modems, scanners, external storage devices, speaker systems, and digital cameras and cell phones. They may work with indus-trial engineers who are concerned with the ergonomics of the devices.

Hardware engineers can be involved in maintenance and repair of com-puters, networks and their peripherals. They troubleshoot problems, order or make new parts and install them. These engineers must be fa-miliar with different network systems as well as programming. Many work as part of a team of specialists who use elements of science, math and electronics to improve existing technology or implement solutions. Computer engineers improve, repair, and implement changes needed to keep up with the demand for faster and more powerful computer sys-tems and complex software programs. What started as a specialty of electrical engineering has developed into a career field of its own.

Hardware engineers are usually responsible for a significant amount of technical writing, including project proposals, progress reports and user manuals for the systems they design.

Software Engineers

Software engineers design computer software programs and customize existing software programs to meet the needs and desires of a particular business or industry. They work very closely with the hardware engineers throughout the design process. They spend considerable time research-ing, defining and analyzing the problem, and then they develop software programs to solve it on the computer. Most large corporations use cus-tomer focus groups to see how their programs are being used. Focus groups help companies make their products easier to use.

There are three types of software engineers, systems software, applica-tions software and realtime software engineers. System software engineers build and maintain the programming infrastructure sometimes called operating systems for entire computer systems while applications software engineers design, create and modify general computer applications software or specialty utility programs based on the system software.

Microsoft Windows is an example of system software whereas Microsoft Word is an example of application software. Realtime software engineers create software that control products that operate in real time, like the computers in your car, or the space shuttle. They write software that control real time processes. Today, most all modern airplanes are controlled by realtime software. AKA as fly by wire, where electrical signals are controlling the moving parts verses flying by direct connection to the moving parts. The days of cables connecting the rudder and other moveable surfaces of a plane to pedals in the cockpit are gone. Software engineers may coordinate customer service, ordering, inventory, billing, shipping, and payroll recordkeeping. They may set up intranets or networks that link computers within the organization to ease the flow of communications.

Application software engineers develop packaged software applications such as word processing, graphic design, spreadsheet, internet, or database programs. Both systems and applications software involves extremely detail oriented work. Since computers do only what they are programmed to do, engineers have to account for every bit of information with a programming command. Software engineers are thus required to be very well organized and precise. Virtually all of their time is spent sitting in front of a computer themselves typing in the instructions that make up the whole program.

Like Hardware Engineers, Software engineers are usually responsible for a significant amount of technical writing, including project proposals, progress reports and user manuals. They are required to meet regularly with clients in order to maintain project goals. When the program is completed the software engineer gives a demonstration of the final product. Sometimes they may install the program, train users and make arrangements for ongoing technical support.

History of Computer Engineering
The first electronic computer, (1943) is the Colossus.

By designing a huge machine now generally regarded as the world's first programmable electronic computer, the then British Post Office Research Branch played a crucial but secret role in helping to win the Second World War. The purpose of Colossus was to decipher messages that came in on a German cipher machine, called the Lorenz SZ also known as Enigma.

The original Colossus used a vast array of telephone exchange parts together with 1,500 electronic valves (vacuum tubes) and was the size of a small room, weighing around a ton. This 'string and sealing wax affair' could process 5,000 characters a second to run through the many millions of possible settings for the code wheels on the Lorenz system in hours rather than weeks. During the succeeding decades, computers became faster and more versatile: the industry was dominated by IBM. Big Iron, as these mainframe computers are called. They were the most common application of electronic computers until the invention of the mini computers and personal computers.

Nearly thirty years later in 1972 the first desktop all-in-one computer, which are more familiar to us today was made. That honor falls to Hewlett Packard with the HP9830, it was the size of an office desk.

The first major advances in modern computer technology were made during WWII. The introduction of semiconductors made possible smaller and less expensive computers. Advances in computer technology have put computers to work in a variety of areas that were once thought impossible. Computer software engineers have been able to take advantage of computer hardware improvements in speed, memory capacity, reliability, and accuracy to create programs that do just about anything.

Environmental Engineers
Environmental engineers design, build, and maintain systems to control waste produced by municipalities or private industry. Such waste streams may be wastewater, solid waste, hazardous waste, or contaminated emissions to the atmosphere (air pollution).

It is the job of environmental engineers to investigate and design systems to make the water safe for the flora and fauna that depend on it for survival. Environmental engineers investigate problems stemming from systems that aren't functioning properly. They have knowledge about waste water treatment systems and have the authority to enforce environmental regulations.

Once a problem has been identified the environmental engineer and the officials at the source of the problem work together on a solution. Some environmental engineers are hired by companies to help keep the company in compliance with federal and state regulations while balancing the economic concerns of the company. Some companies rely on environmental engineering consulting firms instead of keeping an engineer on staff.

Consulting firms usually provide teams that visit the plant, assess the problems, and design the systems to get the plant back into compliance.

Broadly speaking environmental engineers may focus on three areas: air, land and water. Air includes air pollution control, air quality management, and other specialties involved in systems to treat emissions. The private sector tends to have the majority of these jobs. Land includes landfill professionals, for whom environmental engineering and public health are important areas. Water involves activities such as water pollution.

Environmental engineers spend a lot to time on paperwork, writing reports reviewing studies, interpreting regulations, developing policy, etc. They also might climb a smokestack, wade in a creek, or go head on with a district attorney in a battle over a compliance matter. Occasionally there are contradictory regulations that apply to the same area of environmental controls. This can delay a solution or even cause frustration in trying to solve a problem. If they work as a consultant they may have to balance the economic costs of prevention or remediation with still meeting the requirements of the regulations.

History of Environmental Engineering

For several thousand years, up until the mid-1800's, the human population was concerned only with what it could get from the environment.

Mother Nature was forced to care for herself as best she could. The practice of garbage collection, in some early settlements, was to throw the household trash into the street in front of the home, and let roaming bands of pigs take care of it. Water was obtained from a central well within the community, with no safeguards on the quality of the water. Sanitation like this was common, from shear ignorance, and as a result, diseases such as cholera and plague.

It is only within the past 50 years that a separate professional category has been recognized for environmental engineers. The beginning of the sensitivity of the people to the environment is traced to Rachel Carlson's book, Silent Spring . Post civil war industrialization and urbanization created life threatening water and air quality problems. These problems continued during and after World War II. After the war, pollution control technologies were developed to deal with the damage. In the 1930s through the 1960s, someone who wanted to be an environmental engineer would have been steered toward sanitary engineering which basically deals with things like water waste and putting sewers in place. Sanitary engineering is a form of civil engineering.

Environmental engineering is an offshoot of civil engineering. Sanitary engineering focuses on the development of the physical systems needed to control waste systems. Civil engineers who already were doing this type of work began referring to themselves as environmental engineers around 1970 with the great boom in new environmental awareness and the regulations that followed.

Industrial Engineers

Industrial Engineers design the aesthetic look and feel of products from large machines to automobiles to home appliances. Industrial Engineers also design manufacturing and industrial processes for the workers who make products. They are primarily Mechanical Engineers with a focus on product packaging, product aesthetics and how workers use the industrial processes used to make products. They can also be studious, intelligent, glorified accountants, not necessarily processing any significant mechanical abilities or interests in machine design.. This is a good type of engineering for the average engineer not for the super engineer type that eats and sleeps for engineering.

Beginning in the 1970s they also became involved in ergonomics when the public became aware that their use of certain products were causing physical injury, backaches from certain types of chairs or carpal tunnel syndrome from typing, just to name two. Industrial Engineers use focus groups and gather other types of user feedback to determine the acceptance of their designs.

These engineers evaluate ergonomic issues which involve the relationship between human capabilities and the physical environment in which they work. They might evaluate whether machines are causing physical harm or discomfort to workers or whether the machines could be designed differently to enable workers to be more productive.

Industrial Engineers use the principles of Fredrick Taylor's time and mooting studies to predict the labor effort and time needed to make a product.

Materials Science Engineering

Materials science engineers design, fabricate and test materials. They may work to make automobiles lighter and more fuel efficient by creating stronger and lighter metals or by using composite materials. They may help to create artificial knees, hips and elbows using special polymers or they may design new materials for the next space shuttle. They can work with any type of materials; plastic, wood, ceramic, petroleum or metals and create completely new synthetic products by rearranging the molecular structure of otherwise common materials.

Material science engineers have many career opportunities leading to careers in transportation and communications as well as other areas. The percentage of materials science engineers in the total of all engineers in the engineering profession is small, however the need to apply basic materials principles to the solution of engineering problems is great.

The most common academic department for materials field is materials science and engineering. Many universities have departments of metallurgical, ceramics or polymer engineering also. Materials science tends to refer more to the analysis of materials and development of new ones as well as the manufacturing or working with materials. You will have a lot of chemistry combined with metallurgy, ceramics and polymers in your studies. Not all engineering departments will have a four year under-graduate program. Five year programs are common.

History of Material Science Engineering

There has always been the medieval fantasy of turning something into gold, usually lead. We hear this theme in numerous stories. It is what is called alchemy. Mining and metal refining developed gradually over the centuries, spurred by the discoveries of new metals, the desire for precious metals (gold and silver) and the needs of developing technologies. The technology to extract and refine ore from the ground set the stage for additional technological development to improve materials important in manufacturing and construction. Mining engineers also played an important role in this process This evolution accelerated during the twentieth century most notably in the development of plastic polymers and other synthetic materials. By the middle of the century a materials revolution was well on its way with a host for new materials entering the marketplace. It is interesting to note that it did become technically possible to turn lead into gold although not at a competitive cost.

The last century discovered polymers and plastics. As polymer use increased they offered a lower cost substitute for glass and metal. Then there was the discovery of "warm" superconductors. Perhaps the most exciting aspect of materials science engineering is the replacement of body parts with synthetic materials.

Mechanical Engineers

Mechanical engineers plan and design tools, engines, machines and other mechanical systems that produce, transmit, or use power. Their work varies and they may work in design, instrumentation, testing, robotics, transportation or bioengineering. These engineers try to make other people's jobs better. This is the largest and one of the hardest of the engineering disciplines.

The job of a mechanical engineer begins with research and development. A research engineer explores the project's theoretical, mechanical and material problems. The engineer may perform experiments to gather necessary data to acquire new knowledge. Often an experimental device or system is developed.

The mechanical design engineer takes information gained from research and development and uses it to design commercially useful products. Since the introduction of sophisticated software programs, mechanical engineers have increasingly used computers in the design process. After the product has been designed and a prototype developed, the product is analyzed by testing engineers. Design and testing engineers continue to work together until the product meets the necessary criteria. Once the final design is set, it is the job of the manufacturing engineer or industrial engineer to come up with the most time and cost efficient way of making the product without sacrificing quality.

Some types of mechanical systems, from factory machinery to nuclear power plants, are so sophisticated that mechanical engineers are needed for operation and ongoing maintenance. Mechanical engineers also work in marketing, sales and administration. In small companies they may need to perform many tasks. Some tasks may be given to consulting engineers who are either self employed or work for consulting firms.

Metallurgical Engineers

Metallurgy engineers develop new types of metal alloys and adapt existing materials to new uses. They also work with the atomic and molecular structure of materials in controlled experimental environments, selecting materials with desirable mechanical, electrical, magnetic, chemical and thermal properties.

They may be referred to as metallurgists. There are three basic categories in which they work extractive metallurgists, physical metallurgists and mechanical metallurgical engineers.

Extractive metallurgists separate metals from ores and reclaim materials from solid wastes for recycling. They may supervise and control the concentration and refining processes in commercial mining operations. Sometimes they are involved in the design of treatment plants and refineries. Because minerals are becoming depleted in our environment, extractive metallurgical engineers are always searching for new ways to recycle metals and for more efficient methods for extracting metal from lower yield ores. Extractive metallurgists work with environmental engineers to control the pollution effects of extraction.

Physical metallurgists focus on the scientific side of the relationship between the structure and properties of metals and devises used for metals. They begin their jobs when the metals have been extracted and make them useful. They work on the metals to determine their physical structure and test them for impurities and defects to determine where they can be used.

Mechanical metallurgical engineers take metals and by melting and casting, produce forms that will be sold for a multitude of uses such as automotive parts, satellite components, etc. The field of process metallurgy is quite broad and involves many methods for finishing metals to produce commercially standard products.

History of Metallurgical Engineering

Metals weren't examined until the 19th century, but the roots of the science of metallurgy were developed more than 6,000 years ago. As far back as the stone age when tools and weapons were being made from rocks people discovered that some rocks were better than others for certain things. That's where we get the names, Bronze Age and Iron Age from.

By 4300 BC, metals were being melted and molded into usable forms. People then discovered that metals could be improved by mixing them with other components. From copper came bronze and brass and from iron came steel and stainless steel. Metallurgical discoveries like this helped shape the flow of human civilization.

Physical metallurgy as a modern science dates back to 1890 when a group of metallurgists began the study of alloys. Enormous advances were made in the 20th century. In recent years metallurgy scientists have extended their research into nonmetallic material such as ceramics, glass, plastics and semiconductors.

Mining Engineers

Mining engineers work with minerals and mineral deposits from the earth. These include metals and nonmetallic minerals like coal. Mining engineers conduce preliminary surveys of mineral deposits and examine them to ascertain whether they can be extracted efficiently and economically using either underground of surface mining methods. They plan and design the development of mine shafts and tunnels. They supervise all mining operations and are responsible for mine safety. Mining engineers specialize in design, research and development, or production.

These engineers make the decision as to whether a newly discovered mineral deposit is worth mining. They review geological data, product marketing information and government requirements for permits, public hearings and environmental protection. Based on this review, they prepare rough cost estimates and economic analyses. If it appears possible to mine the deposit at a competitive price and with an acceptable return on investment, mining engineers continue.

Mining engineers supervise the mining operation; they train crews, and do inspections to ensure continued safety. They also monitor the quality of the air in the mine to insure proper ventilation and they inspect the repair mining equipment. Some mining engineers specialize in designing equipment used to excavate and operate mines. Mining engineers also work for firms that sell mining supplies and equipment. Experienced mining engineers teach in colleges and universities and serve as independent consultants to industry and government.

History of Mining Engineering

The development of mining technology stems back some 50,000 years to the period when people began digging pits and stripping surfaces in search of stone and flint for tools. Between 8000 and 3000BC the search for quality flint led people to sink shafts and drive galleries into limestone deposits.

By about 1300 BC the Egyptians and other Near Eastern peoples were mining copper and gold by driving near horizontal entry tunnels into hillsides then sinking inclined shafts from which they dug extensive galleries. They supported the gallery roofs with pillars of uncut ore or wooden timbers.

Ancient Roman engineers made further advances. They mined more ambitiously, sometimes exploiting as many as four levels by means of deeply connected shafts. Careful planning enabled them to drive complicated networks of exploratory galleries at various depths. Mining engineering advanced little from Roman times until the 11th century. From this period on basic mining operations such as drainage, ventilation and hoisting underwent increasing mechanization.

During the 18th century, engineers developed cheap reliable steam powered pumps to remove water in mines, steam powered windlasses, power drills for making shot holes for rock breaking explosives, and powered revolving wheel cutters. After 1900, electric locomotives, conveyor belts and large capacity rubber tired vehicles came into use for haulage.

Nuclear Engineers

Nuclear engineers are involved with accessing, using, and controlling the energy released when nuclei of an atom are split. The process of splitting atoms, call fission, produces a nuclear reaction which creates an enormous amount of nuclear energy and radiation. Nuclear energy and its associated radiation have many uses. Some engineers design, develop and operate nuclear power plants. Others develop nuclear weapons, medical uses for radioactive materials, and disposal facilities for radioactive waste.

These engineers are involved in various types of use and maintenance of nuclear energy and safe disposal of its waste. They work in research and development, design, fuel management, safety analysis, operation and testing, sales and education. Nuclear engineering is dominated by the power industry and the military. Some work for companies that manufacture nuclear reactors, or work directly with public electric utility companies. Some engineers work in food supply using nuclear radiation for pasteurization, sterilization, insect pest control and fertilization. Furthermore they can work conducting genetic research on improving various food strains and their resistance to harmful elements. In the medical field they can design and construct equipment for diagnosing and treating illnesses. They can perform research on radioisotopes which are used in pacemakers, x-ray equipment and for sterilizing of instruments.

Nuclear health physical engineers, nuclear criticality specialists, safety engineers and radiation protection technicians conduct research and training programs designed to protect plant and laboratory employees against radiation hazards. Nuclear fuel research engineers and nuclear fuel reclamation engineers work within reprocessing systems for atomic fuels. Accelerator operators coordinate the operation of equipment used in experiments on subatomic particles and work with photographs, produced by particle detectors of atomic collisions. Many aspects of nuclear engineering are regulated by government laws and agencies.

History of Nuclear Engineering

Nuclear engineering as a science is very young; some of its theoretical foundation was formed with the ancient Greeks. In the fifth century BC, Greek philosophers postulated that the building blocks of all matter were indestructible elements, which they named atoms, meaning indivisible. This atomic theory was recognized over the centuries.

It was revised in by John Dalton (1766-1844). In the following century scientific and mathematical experimentation led to the formation of modern atomic and nuclear theory.

Modern nuclear engineering is traced to 1942 when physicist Enrico Fermi (1901-1954) and his colleagues produced the first self sustained nuclear chain reaction in the first nuclear reactor ever built in Los Alamos, New Mexico.

In spite of the controversy over the risks involved with atomic power, it continues to be used around the world. Most of the growth in the nuclear industry has been in electric energy generation, and mostly in Europe. This is a giant growth industry which will suddenly arise from nowhere and these engineers will be in demand, particularly with security clearances. Nuclear energy is the only easy way for the world to continue living at the standard of living we currently enjoy.

Optical Engineers

Optical engineers apply the concepts of optics to research, design and develop applications in areas of the properties of light and how it interacts with matter. It is a combination of physics and engineering, they study the way light is produced, transmitted, detected, and measured to determine ways it can be used to build devices using optical technology.

Optical engineers can work in three areas of optics; physical optics which is concerned with the wave properties of light, quantum optics which is the study of photons or individual particles of light and geometrical optics which involves optical instruments used to detect and measure light. Other subfields of optics include integrated optics, nonlinear optics, electron optics, magneto optics and space optics.

Optical engineers combine their knowledge of optics with other engineering concepts, such as mechanical engineering and electrical engineering to determine applications and build devices using optics technology. Optical engineers design precision optical systems for cameras, telescopes, or lens systems. They also design and develop circuitry and components for devices that use optical technology such as inspection instruments.

Some optical engineers specialize in lasers and fiber optics. These fiber optic engineers, design, develop, modify, and build equipment and components that utilize lasers and fiber optic technology. They may work with fiber optic imaging, communications allowing voice, data, sound and images to be transmitted over cable, or measurement of temperature, pressure force and other physical features. Virtually all of our telephone and computer network communications is carried over optical cables. The invention of optical cables, fiber optics, was a result of collaboration among optical engineers who defined the optical properties of glass fiber, materials engineers who developed the specific glasses used and electrical engineers who designed the electronics to convert telephone and computer data into light.

History of Optics Engineering

The study of the properties of light began during the 1600s when Galileo built telescopes to observe the planets and stars. Scientists such as Sir Isaac Newton, conducted experiments and studies that contributed to the understanding of light and how it operates. By the mid 1800s, scientists were able to measure the speed of light and had developed means to show how the color bands of the light spectrum shining through a prism were created by atoms of chemical elements. In 1864, James C Maxwell proposed the electromagnetic theory of light.

Optical communication systems date back two centuries, to the "optical telegraph" that French engineer Claude Chappe invented in the 1790s. By the mid-19th century it was replaced by the electric telegraph. Alexander Graham Bell patented an optical telephone system, which he called the Photophone, in 1880, but his earlier invention, the telephone, proved far more practical.

In the 1840s, Swiss physicist Daniel Collodon and French physicist Jacques Babinet showed that light could be guided along jets of water for fountain displays. British physicist John Tyndall popularized light guiding in a demonstration he first used in 1854, guiding light in a jet of water flowing from a tank. By the turn of the century, inventors realized that bent quartz rods could carry light, and patented them as dental illuminators. By the 1940s, many doctors used illuminated Plexiglas tongue depressors.

During the 1920s, John Logie Baird in England and Clarence W. Hansell in the United States patented the idea of using arrays of hollow pipes or transparent rods to transmit images for television or facsimile systems. However, the first person known to have demonstrated image transmission through a bundle of optical fibers was Heinrich Lamm.

The laser was invented in 1960 by Theodore Maiman.

Two of the most important discoveries of the 20th century were the development of lasers and fiber optics.

Packaging Engineers

Packaging engineers develop containers for all types of goods, such as food, clothing, medicine, housewares, toys, electronics, appliances, and computers. In developing these containers engineer's responsibilities include product and cost analysis, breakage, spillage, and spoilage, management of personnel, developing and operating packaging filling lines and negotiating sales. Packaging engineers may also select, design and develop the machinery used for packaging.

These engineers plan, design, develop and produce containers for all types of products. They first determine the purpose of the package and the needs of the end user and the client. The packaging may be for shipping, storage, display or protection. The cost of production and implementation is also considered, as well as the packaging materials long term interaction with the environment.

After a packaging sample is approved, packaging engineers may supervise the testing of the package. Packages are dropped, shaken, kicked and thermally stressed to ensure their durability. Design and marketing factors also are considered by the engineer. Packaging engineers may work with graphic designs and industrial designers. For this, knowledge marketing, design and advertising are essential.

History of Packaging Engineering

The use of flexible packaging materials began with the Chinese; they used sheets of treated mulberry bark to wrap foods as early as the first or second century B.C. During the following centuries, the Chinese also developed and refined the techniques of paper making. The first paperboard carton – often called a cardboard box —was produced in England in 1817. An important step for the use of paper in packaging came with the development of paper bags. Commercial paper bags were first manufactured in Bristol, England, in 1844.

Although glass-making began in 7000 B.C. as an offshoot of pottery, it was first industrialized in Egypt in 1500 B.C. While other packaging products, such as metals and plastics, were gaining popularity in the 1970s, packaging in glass tended to be reserved for high value products. Ancient boxes and cups, made from silver and gold, were much too valuable for common use. Metal did not become a common packaging material until less expensive metals, stronger alloys, thinner gauges and coatings were eventually developed. One of the 'new metals' that allowed metal to be used in packaging was tin. The process of tin plating was discovered in Bohemia in 1200 A.D., and cans of iron coated with tin were known in Bavaria as early as the 14th century. However, the plating process was a closely guarded secret until the 1600s. Thanks to the Duke of Saxony, who stole the technique, it progressed across Europe to France and the United Kingdom by the early 19th century. After William Underwood transferred the process to the United States via Boston, steel replaced iron, which improved both output and quality. The safe preservation of foods in metal containers was finally realized in France in the early 1800s. In 1809, General Napoleon Bonaparte offered 12,000 francs to anyone who could preserve food for his army. Nicholas Appert, a Parisian chef and confectioner, found that food sealed in tin containers and sterilized by boiling could be preserved for long periods. A year later (1810), Peter Durand of Britain received a patent for tinplate after devising the sealed cylindrical can. After it became economical to extract aluminum from ore, tin cans gave way to more modern and lighter weight aluminum.

Plastic is the newest packaging material in comparison with metal, glass, and paper. Although discovered in the 19th century, most plastics were reserved for military and wartime use. Several plastics were discovered in the nineteenth century: styrene in 1831, vinyl chloride in 1835, and celluloid in the late 1860s. However, none of these materials became practical for packaging until the twentieth century. One of the most commonly used plastics is polyethylene terephthalate (PETE). This material only became available for containers during the last two decades with its use for beverages entering the market in 1977. By 1980, foods and other hot-fill products such as jams could also be packaged in PETE.

Government agencies, manufacturers, and designers are constantly trying to improve packaging so that it is more convenient, safe, durable, light weight and affordable.

Plastics Engineers

Plastics engineers work in a wide variety of areas in manufacturing processes. Plastics engineers can develop everything from the initial parts design to the processes and automation required to produce the finish product. They may develop ways to produce clear durable plastics to replace glass in areas where glass cannot be used. Or they may design and manufacture light weight parts for aircraft and automobiles, or create new plastics to replace metallic or wood parts. Plastics engineers may develop less expensive, resistant plastics for the use in construction, or they may develop new types of biodegradable molecules that reduce pollution and increase recyclables.

Plastics engineers' duties include making sure the process is consistent for the creation of accurate and precise parts. They use computers to calculate parts weight and cycle times to monitor the process on the molding press. They help customers with parts design and molds for plastic designs. They take an idea and make it profitable for their company while still satisfying the needs of the client.

History of Plastics Engineering

Thermoplastics were discovered in France in 1828. In the United States in 1869 John Wesley Hyatt created celluloid. His invention brought about the use of celluloid in a multitude of products. In 1909 Leo H. Baekeland produced the first synthetic plastic. This replaced natural rubber. Plastic materials have been developed continuously since then, the greatest coming during WWII. Today, plastics is a major industry that plays a vital role in many industries and activities around the world.

Plastics engineers apply their skills to a large number of professional fields such as the medical field in developing artificial hearts, replacement limbs, implantable eye lenses etc. Where once household products were packaged in glass or metal, most are now packaged in plastic.

Process Engineers

Process Engineers' work includes system engineering, management science, operations research, and ergonomics to determine the most efficient and cost effective methods for industrial production. These engineers are responsible for designing systems that integrate materials, equipment, information and people in the overall production process.

They are involved with the development and implementation of systems and procedures that are utilized by many industries and businesses.

In general, they figure out the most effective ways to use the three basic elements of any company: people, facilities and equipment. Their main focus is in manufacturing. They are interested in the process technology which includes design and layout of machinery and the organization of workers who implement the tasks. They also develop methods to accomplish production tasks such as the organization of an assembly line. In addition they devise systems for quality control, distribution and inventory. A process engineer may design the entire product flow within a large production floor.

In industries that do not focus on manufacturing, process engineers are often called management analysts or management engineers. In the healthcare industry, such engineers are asked to evaluate current administrative procedures. They are called upon to streamline the flow of paper and materials. They also advise on job standards cost containment and operations consolidation. Some Process engineers are employed by financial services companies.

Process engineers may study an organizational chart and other information about a project and then determine the functions and responsibilities of workers. They devise and implement job evaluation procedures as well as articulate labor-utilization standards for workers.

History of Process Engineering

The roots of process engineering can be traced to the ancient Greece where records indicate that manufacturing labor was divided among people having specialized skills. The most significant advance in process engineering occurred in the 18th century when a number of inventions were created for the textile industry. By the late 18th century the industrial revolution was in full swing. Innovations in manufacturing were made, standardization, creating interchangeable parts was implemented and specialization of labor was increasingly put into practice.

Process engineering as a science is said to have started with the work of Frederick Taylor in 1881. He began to study the way production workers use their time. He introduced the concept of time study, whereby workers were timed with a stopwatch and their production was evaluated. He used the studies to design methods and equipment that allowed tasks to be done more efficiently.

In the early 1900s the field was known as scientific management. Frank and Lillian Gilbreth were influential with their motion studies of workers performing various tasks. Then around 1913 automaker Henry ford started conveyor belt assembly lines in his factory which led to increasingly integrated production lines.

Process engineers are called upon to solve even more complex operating problems today and design systems involving large numbers of workers, complicated equipment and vast amounts of information. They use advanced computers and software to design complex mathematical models and other simulations.

Quality Control Engineers

Quality Control engineers select the best techniques for a specific process or method, then determine the level of quality needed and take the necessary action to maintain or improve performance. They are responsible for developing, implementing, and directing processes and practices that result in the desired level of quality for manufactured parts. Not only do they focus on ensuring quality during the production process, they also get involved in product design and product evaluation. They will spend a lot of time doing MTBF (Mean Time between Failure) studies and yield analyses to determine acceptable failure rates of products. Only after these studies are completed can a company determine what their warranty and return policies should be. An MTBF study for something as common as a computer motherboard may take as much as 40 hours of intensive labor to perform, most of which is done on the computer. MTBF studies help companies determine warranty times for their products. Quality control engineers may be specialists who work with industrial designers during the design process or they may work with sales and marketing to evaluate reports from consumers. They are responsible for ensuring that all incoming materials meet required standards and that perform properly. They may supervise workers including quality control technicians, inspectors and related personnel. They may also record and evaluate test data.

History of Quality Control Engineering

Quality control is an outgrowth of the Industrial Revolution. It began in England in the 18th century, and led to increases in production and efficiency especially as the manufacturing processes became mechanized during the early part of the 20th century. The mathematical basis for Statistical Quality control was presented by Walter Shewhart in 1932 which provided the basic tools for performing many of the basic inspection techniques still used today. These techniques led to one worker who was responsible for the overall quality of the production. This responsibility belonged to the mechanical engineer until after WWII when a new field of quality control engineer was introduced.

At first they were responsible for random checks of products to ensure they met all standards. During the 1980s the quality movement spread and companies sought to improve quality and productivity. A new philosophy emerged emphasizing quality concerns and that quality be monitored at all stages of manufacturing. Quality control engineers work with employees from all departments to train them in the best quality practices and to seek improvements to manufacturing processes to further improve quality levels. Quality engineers work closely with manufacturing , marketing , legal and government agencies to ensure that products meet the requirements of the customer and of the agencies that regulate industry such as the USFDA, or the Nuclear Regulatory Commission or the FCC or the SEC.

Robotics Engineers

Robotics engineers design, develop and program robots and robotic devices, including peripheral equipment and computer software. The majority of robotics engineers work within the field of computer integrating manufacturing or programmable automation. In many cases these engineers may have been trained as mechanical, electronic, computer or manufacturing engineers.

These engineers have a strong knowledge of computers and an understanding of manufacturing production requirements and how robots can best be used in automated systems to achieve cost efficiency, productivity and quality. Robotics engineers may analyze and evaluate a manufacturer's operating system to determine whether robots can be used efficiently. Taking the human element out of the manufacturing process also makes the products themselves more uniform.

History of Robotic Engineering

The idea of robots can be traced back to the ancient Greek and Egyptian civilizations. An inventor from the first century AD, Hero of Alexandria invented a machine that would automatically open the doors of a temple when the priest lit a fire on the altar. During the later periods of the Middle Ages, the Renaissance and the 17th and 18th centuries, interest in robot like mechanisms turned mostly to automatons, devices that imitated human and animal behavior. The Industrial Revolution inspired the invention of many different kinds of automatic machinery.

The word robot and the concepts associated with it were first introduced in the early 1920's. In 1954 George Devol designed the first programmable robot in the United States. He named his company Universal Automation which became the first robot company. Hydraulic robots were developed in the 1960s and were used initially by the automobile industry in assembly line operations. By 1973, robots were being built with electric power and electronic controls which allowed greater flexibility and increased uses.

Building robots involves the development of a wide range of skills, including creative thinking, design, mechanics, electronics and programming, all of which are highly valued in industry. Your interest in the subject could lead you into an exciting and fulfilling career at the cutting edge of technology!

Today robots are used in all phases of industry and are finding application in home, office and in tasks that are dangerous, such as bomb disposal and handling nuclear waste or physically demanding such as welding.

Traffic Engineers

Traffic engineers use information they gather to design and implement plans and electronic systems that improve the flow of traffic and safety on our roads. They study factors such as signal timing, traffic flow, accident probabilities, lighting, road capacity, and entrances and exits in order to increase traffic safety. They may conduct studies and implement plans to reduce the number of accidents or they may be asked to prepare traffic impact studies for new residential or industrial developments.

History of Traffic Engineering

During the early colonial days, dirt roads and Native American trails were the primary means of land travel. In 1806 the US congress provided for the construction of the first road known as the Cumberland Road. More and more roads were built and land travel became more complex. Traffic engineers were trained to ensure safe travel on roads and highways.

Systems Engineers

Systems engineering grew out of a need for specialized engineering disciplines to be able to work together. They focus on making different components, produced by specialty engineers, work as a cohesive and efficient system or product. Systems engineers also repair and improve upon existing systems as new technologies emerge.

A college degree in systems engineering provides the necessary skills needed to interact with professionals engaged in a broad field of disciplines. Systems engineers must think holistically, taking into account every aspect of a project, including the costs involved, environmental concerns, timeframes, and life expectancy of equipment.

The demand for systems engineers is on the rise, as all kinds of systems become increasingly complex and companies' needs cannot be met by engineers concentrating in a specific discipline, such as electronics, manufacturing, or computers. The birth of a global economy has also stimulated the need for systems engineers, as foreign and domestic platforms are often incompatible.

Bringing together engineers, businessmen, and employees from all parts of the globe and all walks of life is central to a career in systems engineering. Communication and mediation skills are a necessity in the field, as is the need to think of the big picture, instead of getting caught up in the details.

Professionals in the systems engineering field must be able to trust that the engineers involved in a project are doing their jobs correctly. They should be able to manage the creation of a system without getting deeply involved at every level, as the schedules of most projects do not allow for systems engineers to know every last detail of what is happening all of the time.

Systems engineers must also always be looking toward the future. This means that, when developing a system, they must be thinking of ways to improve upon the system in the future with minimal cost and time spent. Since technology is constantly growing and changing, developing systems that are easily adaptable to new technologies can decide the fate of a company. Thus, the responsibility of the systems engineer is very great indeed.

The most important aspects of systems engineering include:
- Design compatibility
- Definition of requirements
- Management of projects
- Cost analysis
- Scheduling
- Possible maintenance needs
- Ease of operation
- Future systems upgrades
- Communication among engineers, managers, suppliers, and customers in regards to the system's operation

A systems engineering major not only prepares the student for a career in systems engineering, but provides engineers with a deeper understanding of how systems work should they decide to specialize in a particular specialty later on down the road. This makes communicating with other systems engineers, project managers, customers and suppliers much easier. The study of systems engineering is all about the coordination of a team, and companies are more likely to hire someone with experience in team coordination.

The systems engineering major covers a broad range of disciplines, providing insight into everything that goes into the production of a process or product, including:
- Physics
- Advanced mathematics, including calculus and differential equations
- Biology
- Chemistry
- Computer science
- Materials
- Design engineering

- Civil engineering
- Business ethics and management
- Writing and communications

To be a successful systems engineer, you should be a natural problem solver and excellent communicator. You need to be able to consider multiple factors and figure out ways for all of these factors to come together and form a whole process. This is called having a "systems view" and comes in handy no matter what field you decide to go into.

Systems engineers must wear many hats, serving in turn as the leader, the listener, the negotiator, and the diplomat. Careers in this field require a great deal of patience and resourcefulness, as well as good old-fashioned common sense. A desire for success without compromising ethics and ideals, as well as a keen understanding of teamwork, are necessary for a career in systems engineering.

Not to overlook the technical side of things, systems engineers must be mechanically and technically competent. Their math and science skills must be top-notch. They must manage their time in the most effective way possible, making sure that every aspect of a project is given careful consideration.
Lastly, systems engineers must possess the desire and ability to continue learning throughout their careers. As technology changes, you must keep up. Survival and advancement in the engineering world requires adapting to change, learning from mistakes, and embracing new and different ideas.

When preparing to enter the field of systems engineering, it is important to take as many advanced math and science courses as possible. Most engineers start out with a Bachelor of Science in Systems Engineering, offered by many traditional and accredited online colleges and universities. Knowledge of programming languages may also come in handy, though it is by no means a prerequisite. Experience in the field, whether it is working on a production line or being a quality control inspector, is always useful, though not an option for everyone.

History of Systems Engineering

The term systems engineering dates back to Bell Telephone Laboratories in the early 1940s and some of their employees trace the concepts of systems engineering within Bell Labs back to early 1900s and describes major applications of systems engineering during World War II. The first attempt to teach systems engineering as we know it today came in 1950 at MIT by Mr. Gilman, Director of Systems Engineering at Bell.

Systems engineering was defined as a function with five phases: (1) system studies or program planning; (2) exploratory planning, which includes problem definition, selecting objectives, systems synthesis, systems analysis, selecting the best system, and communicating the results; (3) development planning, which repeats phase 2 in more detail; (4) studies during development, which includes the development of parts of the system and the integration and testing of these parts; and (5) current engineering, which is what takes place while the system is operational and being refined.

Structural Engineers

Structural engineering is a branch of civil engineering, and its applications are extremely diverse. A great deal of what structural engineers do involve designing things to be built, and then helping to build them: buildings, bridges, tunnels, towers. The architect comes up with a building design, and then it's the structural engineer's responsibility to fit the structure to the architecture, and decide on what structural system is best suited to that particular building. We design the beams, the columns, the basic members to make the building stand up.

A structural engineer might also be involved in the demolition or dismantling of a structure, either permanently or in order to repair it. For both of these processes, they need to know about the forces that act on structures—the stresses put on a bridge by heavy traffic or on a high building by strong winds, or on any structure by seasonal temperature changes or natural disasters such as hurricanes or earthquakes.

Structural engineers also inspect buildings, both during and after construction, and oversee the use of the concrete, steel and timber of which they are made. They must also be aware of both obvious and inobvious uses for the structures and how these uses affect its design. For example, if they're putting in sensitive computer equipment or doing pharmaceutical work, you have to use a floor system that's very stiff and doesn't move much.

Like all engineers whose work may affect life, health or property, new structural engineers go through a rigorous training process during their first few years of work. This training involves several years of work experience under the supervision of experienced engineers and one or more state examinations, and results in a license as a Professional Engineer (P.E.). This is one profession where an advanced degree is more of a necessity than an option.

My advice to students is that if they're really committed to structural engineering, they should get their master's degree in structural engineering or civil engineering as quickly as possible. The basic courses at the undergraduate level just can't touch on all the necessary aspects of structural engineering. Your advance in the profession is greatly impaired by not having a master's degree.

Along with technical know-how, a structural engineer needs a host of other skills to be able to interact with professional and nonprofessional co-workers and clients. Sales ability, public speaking and time management are very important. Problem resolution is a skill that isn't typically taught in engineering schools. When there's an enormous amount of work that costs a lot of money, that is going on very rapidly, and there are problems, then the problems have to be resolved as quickly as possible.

Structural engineers, like other civil engineers, frequently hold the lives of others in their hand.

History of Structural Engineering

Structural engineering developed as a separate engineering discipline in the second half of the nineteenth century in response to the specialist requirements of the design of tall iron and steel framed buildings and the introduction of reinforced concrete, eagerly employed by adventurous architects who lacked the engineering expertise to design such structures. Before that time the discipline had been slowly developing as mankind built ever more ambitious structures.

The scale of prehistoric structures such as Stonehenge reveals that erection rather than design was the major challenge. That this was mastered is displayed in the scale of the buildings of the Egyptians, Greeks and Romans. The Romans were able to build arches and vaults on a considerable scale, and mastered many building materials such as concrete in a way which was not seen on the same scale Europe for centuries. The medieval master mason and military engineer displayed similar skills, although the lack of written evidence makes it difficult to be sure of the design process in detail. The practical evidence in the form of gothic cathedrals and elaborate fortifications demonstrates that the art of the structural engineer was being practiced under a different name, often by master masons and master carpenters

In the mid eighteenth century these traditional skills were presented with new challenges by the industrial revolution which required larger buildings to house larger and heavier machinery, as well as providing new materials in the form of cheaper cast iron, from the 1750s, and wrought iron, from the 1790s, and concrete, from about 1800.

Conclusion

This chapter has exposed you to many of the types of engineering. It defined the five areas of engineering work: research, development, application, management and maintenance. It also defined the five primary career paths that engineers follow: industry, consulting, government, academics and internet. It enlightened you on the long history of engineers in different cultures.

The six traditional areas of engineering were described: chemical, civil, electrical, industrial, materials science and mechanical and you now know that preparation in any one of these can lead you into the many specialties. You have a better idea of what area you will want to go into and why.

Specialties such as aerospace, agricultural, automotive, biomedical, computer, environmental, manufacturing, and petroleum engineering were just a few that were discussed. You now know what each one entails and its history and what qualities are needed to succeed in that specialty.

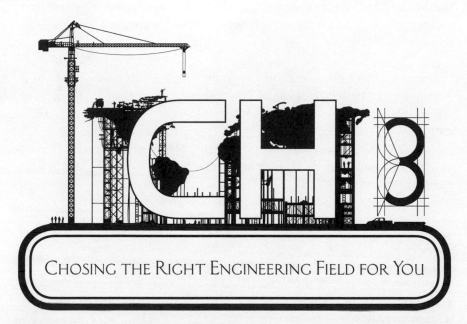

Chosing the Right Engineering Field for You

Since there are many types of engineering fields, you should narrow it down to the field that you are most interested in and that you are most compatible with. First, what are you best at doing and what do you like doing? To find out there are some things you can do such as join the Junior Engineering Technical Society (JETS), participate in a work study program and summer internships, interview engineers, and volunteer in engineering companies. These are a great way to find what type of engineering suits you. Think about the kinds of hobbies you have. Do you fly model planes? Are you a ham radio operator? Did you use your rock collection for your high school science fair? Did you have an erector set or Lego set? Did you have a chemistry set?

How to Find Out What You are Most Interested in Aerospace

If you enjoy building and flying model airplanes and rockets you may be a good fit for aerospace engineering. You may also want to work on cars and boats which will provide good opportunity to discover more about aerodynamics. You should enjoy completing detailed work, problem solving and participating in group efforts. Curiosity, inventiveness and willingness to continue learning from experiences are excellent qualities to have for this and any type of engineering.

Summer camps and academic programs offering studies focusing on aerospace can be very helpful. You can also apply directly to agencies of the federal government concerned with aerospace development. Applications can be made through the Office of Personnel Management. Or join a professional association such as the National Society of Professional Engineers and the American Institute of Aeronautics and Astronautic Engineers. These also offer job placement services. There are general Science Summer camps available to students of all ages.

Biomedical

If you think you might be interested in biomedical engineering try volunteering or working in a hospital where biomedical engineers are employed and can provide you with insight into the field. You should have a strong commitment to learning if you plan to work in this type of engineering. Also being scientifically inclined and being able to apply that knowledge in problem solving is essential. You also have to be familiar with chemical, materials and electrical engineering as well as physiology and computers.

Ceramics

You should be an inquisitive person with somewhat of an analytical mind to pursue Ceramics engineering. You also need to be comfortable working with details and should enjoy doing intellectually demanding work. For hands on experience with materials, take pottery or sculpture classes. This will help you become acquainted with materials you will be working with as a Ceramics Engineer. You might want to inquire at manufacturing companies for internships and summer employment.

Chemical

If you think you might be interested in chemical engineering it is important to have personal qualities such as honesty, accuracy, objectivity and perseverance. You should be heavily interested in advanced math and chemistry. In addition you should be inquisitive, open minded, creative and flexible. Problem solving ability is essential. Also leadership abilities, working well on a team, and collaborating with people of different cultures and speaking different languages is also helpful.

Civil

Basic personal characteristics often found in civil engineers include an avid curiosity, a passion for mathematics and science and an aptitude for problem solving. You should be able to work alone or with a team and visualize multidimensional and spatial relationships. Getting involved in the actual building of a project design will provide experience and exposure to the work of a civil engineer.

Electrical

If you like buildings radios, or are a ham radio operator, or enjoy tinkering with computers you may be a candidate for an electrical engineer. To be a good electrical engineer you should have a strong problem solving ability, strong mathematical and scientific skills and the willingness to learn throughout your career and most important a curiosity for how things work. Often a number of products are worked on simultaneously and you must be able to multitask.

Environmental

A good way to find out if you are interested in environmental engineering is to volunteer for the local chapter of a nonprofit environmental organization or pursue an internship or work first as an environmental technician. Environmental engineers must like solving problems and have a good background in science and math. They must be able to just get in there and figure out what needs to be done and communicate verbally and in writing with a variety of people from both technical and non technical fields. If you have a strong interest in ecology, this may be the field for you.

Industrial

Industrial engineers convey ideas graphically and may need to visualize processes in three dimensions so if you are good in these areas than this type of engineering might be for you. They are also well versed in computer science, creativity and enjoy problem solving and analyzing as well as being a team player. The ability to communicate is vital since they interact with all levels of management and workers. Being organized and detail oriented is important because industrial engineers often handle large project and must bring them in on time and on budget. You may explore your aptitude and interest in an industrial engineering career through graphic design courses, art classes and computer design. If you like model making this may be the field for you.

Mechanical

One of the best ways to find if mechanical engineering is for you is to talk with a mechanical engineer or tour an industrial plant .You might tackle a design or building project to test your aptitude for the field. Personal qualities essential for mechanical engineers include the ability to think analytically, solve problems and to work with abstract ideas. Attention to details is also important and good oral and written communication skills as well as computer literacy are essential.

Metallurgical

Taking sculpturing, foundry, and welding classes is a good way to learn if metallurgical engineering is for you. Creating bronze sculptures, designing and making metal jewelry and welding metals into structures provides hands on experience into this engineering field. You should have a curiosity about how things work and an analytical mind with mechanical abilities. In general metallurgical engineers are interested in nature and the physical sciences and are creative and critical thinkers who enjoy problem solving. They are also patient, well organized, and are attentive to detail because much of their work involves long term projects.

Mining

Certain characteristics help qualify a person for a career in mining engineering. These include the judgment to adapt knowledge to practical purposes, the imagination and analytical skills to solve problems and the capacity to predict the performance and cost of new processes. You have to want to work a good portion of the time outdoors and get close to the earth. If you don't mind getting your hands dirty, this may be the career for you.

Nuclear

If you are interested in nuclear engineering you must accept two major concerns. First is the possibility of exposure to high levels of radiation, thus you must be willing to always follow safety precautions. Also you must be willing to have a lifetime of continuous education. So if you are a very safety conscious person who likes continuous learning this might be the engineering field for you.

Optical

To become an optical engineer you need to be strong in mathematics and physics as well as have an inquisitive and analytical mind. You should be a good problem solver, enjoy challenges and be methodical, precise and attentive to details.

Packaging

Packaging engineers should have the ability to solve problems and think analytically and creatively. They must work well with people, both as a leader and as a team player. They should also be able to write and speak well in order to deal effectively with other workers and customers in order to document procedures and polices. In addition you should be able to manage projects and people.

To get first hand experience in the packaging industry you can call local manufactures to see how they handle and package their products. You should be interested in observing and learning from the packaging of every day items that you come across. You may explore your aptitude and interest in a packaging career through graphic design courses, art classes and computer design.

Plastics

You need to have good mechanical ability in order to develop the plastics parts and the tooling necessary to become a plastics engineer. You must have thorough knowledge of the properties of plastics and when best to use them. You also must be imaginative and creative in order to be able to solve any problems that might arise from new applications or in the transition of a metal part to a plastic one. Many plastics engineers start out as tool and die makers or mold makers before they move into engineering positions. You should be interested in design and manufacturing.

Quality Control

Quality control engineers need scientific and mathematical aptitudes, strong interpersonal skills and leadership abilities. Good judgment is a plus because quality control engineers must weigh all the tradeoffs influencing performance and cost. You should enjoy and do well in math, science and other technical subjects and should feel comfortable using the language and symbols of math and science. You should have good eyesight and good manual skills including the ability to use hand tools. You should be able to follow technical instructions and make good judgments about technical matters. Having an orderly mind and being able to maintain records, conduct inventories and estimate quantities are helpful. You will also need to have good written skills since you may often be called upon to write test procedures and general quality reports.

You will be working with scientific instruments, so therefore you should like academic or industrial arts courses that introduce you to different kinds of scientific or technical equipment if you intend to become a quality control engineer. You should have enjoyed your electrical and machine shop courses, mechanical drawing and chemistry courses. Joining a radio, computer or science club is a good way to gain knowledge to know if this is the type of engineering for you. You should keep in mind that quality control engineers are often involved in manufacturing processes. If it is at all possible, try to get a part time or summer job in a manufacturing setting to get hands on experience.

Robotics

Because the field of robotics is changing rapidly one of the most important requirements that you will need for this type of engineering is the willingness to pursue additional training on an ongoing basis. Robotics engineers need manual dexterity, good hand eye coordination and good mechanical and electrical abilities. If you like learning about the current trends and recent advances in robotics then this field might be for you. You might be experienced in automotive mechanics, machining, or computers.

Computer

Software and hardware engineers need strong communication skills in order to be able to make formal business presentations and interact with people having different levels of computer expertise. You must also be detail oriented and work well under pressure. In general you should be intent on learning as much as possible about computers and computer hardware and software.

Traffic

If you think you are interested in being a traffic engineer you should enjoy the challenge of solving problems and have good oral and written communication skills since you will frequently be working with others. You should also be creative and able to visualize the future working of your designs.

Contractors and Consultants

Contractors and consultants are usually born out of necessity. Nobody begins their career with that in mind. The distinction between the two is usually one of expertise. Consultants tend to be more educated and extremely knowledgeable in a very narrow technical field whereas contractors tend to be generalists and can take on many different types of assignments.

Contracting

An Engineer sometimes becomes a contractor after he or she has been laid off or cannot find a job for whatever reason." Many companies hire contractors. Companies do not like laying off Engineers. It is very costly for them, not to mention the bad publicity and decrease in morale among their regular employees for having to announce a layoff, so when they begin a new project, they may hire contractors to supplement their core engineering team. In this way when the project is over, they merely terminate the contractors.

Contracting is somewhat of a misnomer because there really is not any contract that needs to be signed. In years past, they were called "job shoppers" but that term has fallen out of favor. As a contractor, you can be terminated at any time for any reason, and usually with little or no notice. Although companies may do this to you, it is a very poor idea for you to quit in the middle of an assignment. The company has usually hired you for the duration of a project, or to complete a specific portion of it and would not be happy if you decide to leave in the middle. Even if the working conditions are poor, it is a good idea to continue working there until your time expires.

Most companies will give you some idea of the duration of the project when you first arrive. It may be just a month or it could be for a year or more. Staying on good terms with each company will make you more desirable in the future and it will be more likely that you will be called in again when another new project begins. Building up a good reputation with each company you work for goes a long way toward getting future assignments at those same companies. Many good contractors never have to really go out looking for work. A simple call to their previous companies letting them know you will become available is enough to get an offer to begin a new assignment.

Contractors usually receive their paychecks from a payroll agency. The agency takes care of billing your company and making the necessary deductions from your paycheck for taxes, etc. As a contractor, you will usually negotiate your own hourly pay rate during the interview process which can be anywhere from 50% to 100% more than a regular employee of the company. Most contractors can earn between $120,000 and $200,000 per year.

The payroll agency then tacks on an additional percentage, ranging from 15% to 25% when they bill your company. There is an easy formula for converting an hourly rate to yearly pay. Take your hourly rate, double it and add three zeros, i.e. $40/hour is $80,000/year. Because you are being paid hourly, when you miss a day, you do not get paid. Your payroll agency will supply you with weekly time sheets for you to record your hours. At the end of the week, you would have your manager sign your time sheet and then you would forward a copy to the payroll agency so they can bill your company for your hours.

At large companies that hire many contractors, the payroll agency may have an office onsite. That makes turning in your timesheets very convenient for you. In other cases it may mean mailing it back or physically delivering it to the payroll agency office yourself.

As a contractor, you may need to get your own health insurance. Some of the more progressive payroll agencies have their own benefits programs as a means of attracting contractors. They may offer health insurance, a 401k program and sometimes even sick time pay and vacation allowances. Even this has a downside. If you are between contracts for more than a month, many payroll agencies will terminate your benefits forcing you to go on COBRA until you are payrolled again on your next assignment.

Consolidated Omnibus Budget Reconciliation Act (COBRA) health benefit provisions. It provide continuation of group health coverage that otherwise would be terminated. COBRA contains provisions giving certain former employees, retirees, spouses and dependent children the right to temporary continuation of health coverage at group rates. This coverage, however, is only available in specific instances. Group health coverage for COBRA participants is usually more expensive than health coverage for active employees, since usually the employer formerly paid a part of the premium. It is ordinarily less expensive, though, than individual health coverage.

For you, the upside of being a contractor is that you will earn a lot of money, and you will not be expected to have any allegiance toward the company you are working for. One downside is the lack of job stability. A contractor may find himself working for a different company every few months with periods of unemployment in between. Depending on the economy and the business you are in, those in between periods may also last for months. It is important to have a good handle on your finances. Although you will earn a lot of money while you are working, you will still need to eat and pay the rent during the down time. Having a lot of down time may be a plus for some contractors. Many feel they earn enough money working only six to nine months of the year and use the remaining time to fulfill their personal dreams.

As a contractor, you will not be part of any team where you work. You tend to be left alone. Regular employees will treat you differently, in many cases ignoring you totally except when it comes to technical project issues, and your manager will usually expect more from you than he would from a regular employee. You may not be invited to attend meetings, especially if they are talking about company confidential issues nor will you usually be invited to company functions like holiday parties and other outings. You will also probably find yourself in the least desirable office, with the least number of office amenities. Some companies will not even include you in their telephone directories.

Also, do not expect to be able to go to any technical or other conferences that your company sends their engineers to. Your continuing education will be in your own hands. Taking night classes and going to seminars are an excellent use of your down time between assignments. Technology changes rapidly and the longer you find yourself out of work, the more technology will be passing you by. It is extremely important to keep current in your field. The companies you work for will not be doing it for you.

After being a contractor for any length of time it will be more difficult to go back to regular employment. Many companies have the attitude that if another contract comes along you will quickly jump ship because they are so lucrative, after all, you will be accepting a large pay cut for the stability of regular employment and the benefits that go along with it. Companies view contractors with a lack of trust. Also, having 8 jobs in 2 years does not look good on anyone's resume. The average engineer keeps his regular job between 2 ½ and 3 ½ years.

At many companies, Human Resources' employment specialists are instructed not even to consider a previous contractor for regular employment and will not even forward your resume to the hiring manager. Engineering department managers often feel the same way. "Don't waste my time with an Engineer who's been a contractor." They are often thought of as not being team players and at worst, disciplinary liabilities.

Contracting may also have an adverse affect on your personal credit rating. If you decide you need to borrow money for a home, automobile, or other major purchase, you will not be able to show a bank any long term history of stability in your employment. One way around this is to save money so you can make higher down payments that will allow you to apply for smaller, more easily obtainable loans.

You should not let all this bother you though. Just keep saying to yourself, "I'm earning twice as much as all of them, even my manager".

Consulting

Most technical consultants are truly self employed. Many setup their own one-person corporations and usually do not have any wish to become regularly employed. A good consultant will earn even more than a contractor with many consultants earning upwards of $250,000 per year. Getting a doctorate degree in your field goes a long way toward becoming a top tier consultant. You also tend to be well respected within the company you work for and you will find that other employees will seek out your advice because of your expertise. With the extra money you will earn, you will have to provide for your own medical benefits and pay your own deductibles to the IRS.

In return, you will have total freedom to work or not work as you see fit. If you are good at what you do, companies will be seeking you out and you will be able to pick and choose the assignments you want to take on. It is amazing how fast word gets around. Of course, this assumes good economic times. If business is slow in your field of expertise, you too can find yourself out of work for long periods of time just like any other contractor. If you do consulting for government agencies or have a desire to work on public works projects, you may also consider getting your Professional Engineering (PE) license as it may be a requirement where projects involving public safety are involved.

Your Strengths and Skills

Everyone has unique strengths and skills. It is important to make sure that you are working in the environment you prefer, doing activities and using skills that reflect your natural abilities. If you work with your natural approach you will be satisfied and motivated. You will learn to appreciate how you like to lead or be led and the type of contribution you make as a team member. Knowing your natural work preferences will provide you with a link between work and personal satisfaction.

Your strengths are a combination of education, experiences, values, needs, goals and general genetic makeup. Finding your strengths involves looking inward and asking yourself what can sometimes be difficult questions. Getting a better understanding of your skills can help you evaluate your strengths. One way to identify your skills is to write down a list of all the tasks and responsibilities you had done in any job or volunteer position you ever had. Using this list can point out in your history when you learned how to do something and you can determine if you are at the beginning, intermediate or advanced level in that skill. This will show you where your strengths are and how they may lead you to the right engineering field.

Also, your underdeveloped skills will help you recognize where your weaknesses may be. Recognizing these short comings and planning to overcome them with either on the job training or additional education can be a very positive thing in your life.

Code of Ethics for Engineers

Engineering is an important and learned profession. As members of this profession, engineers are expected to exhibit the highest standards of honesty and integrity. Engineering has a direct and vital impact on the quality of life for all people. Accordingly, the services provided by engineers require honesty, impartiality, fairness, and equity, and must be dedicated to the protection of the public health, safety, and welfare. Engineers must perform under a standard of professional behavior that requires adherence to the highest principles of ethical conduct. Engineers uphold and advance the integrity, honor, and dignity of the engineering profession by:

1. using their knowledge and skill for the enhancement of human welfare;

2. being honest and impartial, and serving with fidelity the publics, their employers;

3. striving to increase the competence and prestige of the engineering profession; and

4. supporting the professional and technical societies of their disciplines.

The Fundamental Canons

1. Engineers shall hold paramount the safety, health, and welfare of the publics in the performance of their professional duties.

2. Engineers shall perform services only in the areas of their competence.

3. Engineers shall issue public statements only in an objective and truthful manner.

4. Engineers shall act in professional matters for each employer or client as faithful agents of trustees, and shall avoid conflicts of interest.

5. Engineers shall build their professional reputation on the merit of their services and shall not compete unfairly with others.

6. Engineers shall act in such a manner as to uphold and enhance the honor, integrity, and dignity of the profession.

7. Engineers shall continue their professional development throughout their careers and shall provide opportunities for the professional development of those engineers under their supervision.

Questions and Answers that will help you Find the Right Engineering Field

Q: I am in high school and want to go into the aerospace field. I am curious about the grades needed to be accepted into college. Some nights I just cannot sleep thinking that I will not be able to pursue my dream when I get older. My grades are mostly in the 80s and low 90s.

A: College admission policies vary considerably from college to college. Among the most important factors colleges and universities consider when evaluating an application are high school grades, difficulty of the high school courses selected, and scores on major standardized tests like the SAT and ACT. Other factors that may play a role include the reputation of the high school, teacher recommendations, the student expressing a strong interest in attending that college, and being the child of an alumnus of that college.

Participation in extracurricular activities is also considered but usually does not have that large an effect. In general, grades in the 80s or higher are good enough for most colleges. Many colleges and universities require a grade point average of at least 2.5 out of 4 for admission.

Q: How many years of school do you need to attend to get a degree for most engineering and scientific fields? How heavy is the workload for these degrees and what classes are the hardest?

A: It takes four years to get a Bachelor of Science degree that is required to obtain a career in most engineering and scientific fields. Another two years or so getting a masters of Science degree, and much longer working on a doctorate. Though generally not required to enter an engineering field, a more advanced degree often improves the quality of positions and salaries offered to an employee.

The course load required for engineering fields in general tends to be one of the most difficult at any university. Engineers generally have to take more credit hours per semester than any other field, and this is one of the primary reasons why engineering students are increasingly needing more than four years to complete a bachelors degree.

The hardest classes will vary for each person depending on his/her talents, the quality of the teaching, and the specific branch of engineering. High level mathematics classes tend to be some of the most difficult, such as Differential Equations or Advanced Calculus. Physics classes probably rank second.

Q: How long would it take for a full time student to get his masters degree and PhD?

A: A masters degree in engineering typically takes one to three years to complete. A doctorate generally requires three to six years, and that is on top of the 4 years required to complete a Bachelors degree. Graduate programs that go directly to a PhD without a masters degree are becoming more common and may shave a year or so off the time.

Q: I've always been interested in aircraft and want to design planes for the military. The problem is that I am disabled due to Muscular Dystrophy. I can still can use the computer and write even though my hands are weak. Can I be an aerospace engineer even though I am disabled?

A: You should not be concerned about your situation interfering with a career as an engineer. Engineers who have a variety of disabilities are still very productive, highly qualified engineers. One of the brightest, hardest working engineers I know is confined to a wheelchair because of Muscular Dystrophy but this has not prevented him from becoming a senior aerodynamicist at a major aerospace company. Other engineers who are paralyzed because of accidents, legally blind, deaf, or missing limbs have all become successful engineers. Most aerospace companies and research facilities are equipped with facilities to support handicapped employees.

Q: I am thinking about majoring in engineering. I like learning about space, but do aeronautical engineers have much to do with space besides designing spacecraft? Should I study engineering or another field?

A: Depending on what it is you enjoy learning about space, I suspect you may prefer a career in science over one in engineering. Scientists like astronomers and astrophysicists study the structure of the universe and the objects within it. These studies often depend on the development and use of new technologies that make observations of distant objects possible. Engineers are the designers who take the requirements specified by scientists, solve the technical problems involved, and create devices to collect the information scientists need.

If you are more interested in designing spacecraft and scientific instruments or calculating trajectories that bring spacecraft to their destinations, you will probably be happy as an engineer. If you prefer studying stars, planets, galaxies, and other celestial objects, you would be better off choosing a career in science. Some possible options include astronomy (the study of celestial objects), astrophysics (the study of the physics of the universe), astrobiology (the study of life in the universe), planetary science (the study of planets and solar systems), and astrochemistry (the study of chemicals in space).

Q :I am an aeronautical engineering student and want to be a rocket scientist. How can I become one?

A: The term "rocket scientist" is actually rather vague and its meaning can vary depending on the context. Rocket scientist is often considered a slang term for anyone with an aerospace engineering degree whether that person actually works on rockets or not. Once you complete your aeronautical engineering degree, you would already be considered a rocket scientist by this broad definition of the term.

Another meaning applies only to people who work on rockets, usually specifically to those working on rocket propulsion systems. These experts analyze propellants and the combustion process or design the components of a rocket propulsion system for use in vehicles like space launchers or military weapons. To specialize in this field as an aeronautical engineering student, you should emphasize your coursework in subjects like fluid dynamics, combustion science, or electrical propulsion. Even this meaning of the term rocket scientist can become broad since not only aerospace engineers work on rocket propulsion technologies. Other types of engineers and scientists can also be considered rocket scientists such as chemists or chemical engineers, material scientists, mechanical engineers, electrical engineers, and physicists.

The important thing to remember is that there is no rocket science degree that a student can earn in college. The term rocket science applies to a variety of different fields required to make advanced vehicles like aircraft and rockets possible.

Q: I am confused about becoming an aerospace engineer or a commercial pilot. I want to know what do aerospace engineers do compared with pilots?

A: There is very little in common between aerospace engineers and commercial pilots. Engineers design vehicles, write software, conduct testing of vehicles and their components, and develop the technologies needed for air and space travel. Pilots fly the vehicles that engineers create. Both careers require highly skilled and trained professionals but the work environment is considerably different.

Q: How long does an engineer have to work in a typical week? Is it hard to balance work and family?

A: Most engineers work a 40-hour week, comparable to any other office career. Overtime can be common when deadlines approach and time to complete a task is limited. Working 60 hours or more per week can be typical at times like these. Most engineers I know seldom have to work more than a normal 40-hour week and they also say one of the best things about their career is the flexibility to customize their work schedule. Some people prefer to start work very early and complete their work day in time to pick up their children from school. Others prefer to stay home with their families during the day and do most of their work later in the evening and at night. So long as their tasks are completed, most employers are willing to let their employees customize their schedule depending on preference or family situation. There are times when an engineer's job and family commitments interfere with each other, but these instances are probably less frequent than in many other careers. In general, engineers are salaried employees meaning that they have to put in whatever number of hours are required to get the job done, satisfactorily, and on time. Companies that offer flextime are an advantage. I always preferred to come to work very early and leave before rush hour.

Q: I am a creative person. I enjoy engineering and I think I may minor in marketing, but I don't know what to major in. I was thinking about going into architectural engineering but I'm not sure. I also have heard rumors that being a women will benefit me if I go into an engineering field. I like math but if there is any field that doesn't require hardcore math everyday please tell me about it.

A: Go into whichever field interests you most. They all require a lot of math; and most of the courses you take will contain a lot math. Being a woman does help you get scholarship money. Rather than minor in Marketing, I might suggest getting an MBA sometime after you begin your engineering career.

Q: I am about to pick between software engineering and database development. By software engineering, I mean designing, coding, testing, project management, and getting more advanced in what most undergraduates do. Databases narrow to relational databases, data mining, data warehousing, query optimization, indexes, and maybe a bit of web applications. I've done pre requirements for both and really enjoy them. What would be the common differences between those paths.

A :Personally, I would recommend taking the software engineering route, and then just learning about databases on the side. I personally think it's much easier to learn about databases by doing. Install a database on your computer, and learn on your own how to make it work. Software engineering is a broad field and will allow you to go into many different areas whereas databases are only one aspect of software development.

Q: Should I leave teaching and go back to Engineering. I have a BS in mechanical engineering and am teaching. Having an engineering degree does not qualify me to teach math, I would have to go back to school and take additional courses. I am considering leaving the teaching profession. My only concern is that I have no engineering experience for the last five years. I am afraid that this will reflect poorly on me when I decide to interview for engineering jobs.

A: If you can get a job in the engineering field do it. You'll make better money. It may be tough since you have no recent experience but look for the things that are common in each field: managing people, your organization skills, initiating communications, maintaining records, etc. Interview for entry level positions, be dynamic and someone will hire you! Once you get that first position, it will become easier to become current in your field.

Q: Which engineering field should I go into? I'm weak in math?

A: Everything in Engineering is math. Any engineering field will require much math.

Your best bet is to get good at math. It. is just like anything else. It can be mastered by anyone who is willing to practice .Even if you struggle with early math (such as algebra), practice enough and you will begin to develop a brain for mathematics. Just like learning a new language, at first even easy words are hard. But once you can speak the basic language you can learn new difficult words much easier. If you have trouble with math and wish to be an engineer then start math classes when you start college. Make it your sole focus to understand the math. This often means lots of homework and practice as well as reading the textbooks (as hard as that is) over and over until you internalize the concepts. As you progress it will get easier and easier. By your second or third year of math, math should become very intuitive and your brain will be fully prepared to take on difficult mathematics with only a little thought.

Q: I am wondering what the highest paid Engineering field is and what would be a good field to take up when I go to college next year. My strong points are in math and sciences particularly in physics and chemistry.

A: Chemical Engineering is the highest paid undergraduate degree

Q: I have completed grade 12. What is the required degree to pursue a masters degree in aerospace/aeronautical engineering in the US? Is it okay to have a bachelors degree in electronics engineering? I want to take up electronics now, to have a greater range of options to choose from later, in case I change my mind about aeronautics later. Assuming I still want to pursue aeronautics 4 years down the line, what can be taken up now(except aeronautical engineering)?

A: Most masters programs in aeronautics in the U.S. require a bachelor's degree in aeronautical or mechanical engineering. Check with your school of choice to be sure — they all have websites.

Q: I am looking at either civil engineering or electrical engineering. I'm mainly considering these to get qualified for sitting for the patent bar that will allow me to concentrate on intellectual property as an attorney. But I'm also looking at architecture as a possible future career.

Which one of those would benefit me the most? I know electrical engineering is more beneficial to a lawyer, since it's a highly desirable qualification to have. But would civil engineering be better if I want to go into architecture one day?

A: I say Electrical. I find it much more interesting than most other fields and, as a general observation (that is, not necessarily my opinion, just what I've seen and heard), one of the most respected engineering fields. But really what you should look at is what you are more likely to do with your degree. If you figure that they're about equally as likely, do what you enjoy more. You'll take a lot of classes in general engineering before you have to decide which specific field, so you've got some wiggle room.

Q: I'm a first year engineering student, but I'm not sure which type of engineering to go into. I thought originally to go into Mechanical because I want to learn more about engines. But I also like chemistry a lot, so I was also looking at Materials Engineering. Any thoughts? I would appreciate any advice.

A: It sounds like you need to talk to a counselor to find out more about each field. Mechanical Engineering is MUCH more than engines. If you like chemistry and mechanical "things" consider looking into chemical engineering.

Your counselors will be your best avenue right now unless you know of more specific interests you have that can help narrow down the field.

Q: I'm in high school, and wondering what fields to enter into for college. I am interested in music and science and I figured Audio Engineering would be a good field to enter to get a little of both. I looked for information on the internet but there's not a lot about it. Can anyone give me some counseling, advice, or even references?

A: Having worked with many engineers (both live sound and in recording studios), and seeing one of my bass player friends develop into the industry's head engineers for over a decade, I know it can be a really satisfying and fun way to pursue both music and technology.

Like many things in the creative arts, the best engineers are in it because it's their passion. It allows you to create and influence creativity at a level that combines some of the most recent emerging technologies, working with incredibly gifted musicians, composers and producers, and maybe even world travel.

But like many other things in the creative arts, pursuing your passion is never an assured path to financial comfort. It's a cliché in every aspect of the music industry that legions of talented people struggle with two or three day jobs while untalented hacks nail the really juicy money gigs because of who they know rather than what they are capable of doing. And engineering (both live sound and studio) are the same.

So just set your expectations accordingly.

I can tell you that most of the formal education programs I've seen exist at the vocational school level, not at the four-year college/university level. There are programs attached to recording studios (like Full Sail Studios), and there are formal programs at music schools like Berkley College of Music in Boston.

As a high school student, I'd suggest you intern or volunteer for a local live sound company or recording studio to see how real life works for engineers. Pick up all the industry publications like Mix, Modern Recording, Electronic Musician, etc. and read interviews with the pros. Get conversant in the basic concepts and vocabulary, etc.

It can be as fun, rewarding and fulfilling as any career in music — and subject to the same liabilities and problems.

Q: I am a senior in high school. For my senior project I am doing a project on Civil Engineering. My questions for the engineers are why is there a lack of women in the profession? Are there specific reasons and could you please expand on the reasons if there are any. Also I need a little bit of advice. For my senior project, I need to have interaction with the class. I need my class to be involved with something hands on during my presentation. Could you suggest a simple way of how I could get my class involved with civil engineering?

A: Regarding why more women don't choose engineering, my thinking is that it's mostly because women before them never chose it. To me, it's like why most men don't go into nursing. It's viewed as a field for the opposite sex, so most people don't even let it enter their considerations when choosing a major in college. It takes a strong, inquisitive mind to fight against these built in stereotypes and ask "why can't I do engineering?" That's what so many women did, and what we try to encourage other women to think. There is no job out there that we can't do, and more of us every year realize that engineering is a valid and enjoyable choice for us. I'm very happy you're looking into the profession!

As for your senior project, there are several different ways you could get your class involved. If your interests lay in highways and traffic, you could present a video of a car crash and have the class divide up into teams to perform the calculations of how fast the car was going and if it was exceeding the speed limit. If you are more interested in structures, you could ask the class to build bridges out of certain materials and vote on whose bridge is most likely to fail under a certain amount of weight, and then take the bridges out and break them. If your interest is more in materials, you could have them design concrete mixes and calculate the slump that each concrete will have, and then take them outside, have them mix the concrete and see whose is most accurate (and if you have more time, you could always have them build concrete cylinders, cure them for 28 days, and then go to your local college and have them broken at the lab to see whose could withstand the most weight).

Q: My family and I have taken several cruises and I absolutely love it. I am amazed at the ship, the design from bow to stern. I love everything about it. I am interested in working on building or designing parts of the cruise ship in some way, shape, or form, but don't know what type of engineering would fall under and more importantly what colleges would offer those types of studies. Can you please give me any input on this or lead me in some sort of direction. Thank you.

A: You might want to look into an Ocean Engineering or Marine Engineering program. These types of programs would include ship design. MIT has an undergraduate program in mechanical and ocean engineering. The US Merchant Marine Academy has a program in Marine Engineering. A mechanical engineering program would also provide the right type of preparation. In that case, I would recommend looking at a program with a string design component in their curriculum. Some of the universities on the coasts may have strong relationships with shipbuilders that may result in interesting student projects.

Q: Hi, I am a fresh high school graduate. I'm planning to study Software Engineering. What other specific fields does my course stretch to? Also, I am having second thoughts about whether I'd survive my course because I rather have a weakness in Math. I have a natural inclination to the arts and music and I'm also very fond of dissecting our computer 'software wise'. I really want to pursue a career in Engineering but I still have doubts.

A: It's great that you are considering continuing your education in a technical field. Earning an undergraduate degree in software engineering (or computer science) means you'll be learning about computers and computational systems, their theory, design, development and application. You will take classes in learning programming logic, different programming languages, algorithms, computer networks and security, computer system architecture, operating systems, database design, etc. There are many different areas you can choose to focus on once you know where your interests are.

Engineering is as much about having a good work ethic and dedication to your studies as it is about applying science and math to the real world. Don't doubt your abilities until you try them out! If you are interested in arts and music, I suggest either minoring in those areas or joining clubs and other activities so you can still have a diverse educational experience and pursue all of your interests. It's very common for students to change their major, so there is no need to worry.

Q: I am currently a senior at Cal State Long Beach and will graduate with my Civil Engineering B.S. in a few semesters. I am interested in structures or transportation (I'm still learning about both, so I can't really say which one I like better at this point). However, I also recently learned about urban planning/development from online. I was wondering if it would be an asset to study urban planning/development as part of a masters program, or if it would make little or no difference. What would be a good match for a master's program for me, if I choose to get a master's degree after getting my B.S. in civil engineering? I am not sure exactly which company I will be working for once I graduate, but I can see myself working for the city or government agencies in possibly structures or transport. I am simply exploring my options and was wondering on your opinions about this. Thank you.

A:. A combination of pure Civil Engineering Design with a Roadway Design focus and a masters in Urban Planning / Development would be a good complementary fit for the future direction of the civil industry. Let me elaborate on that. When we look at how we will build things in the future, the trend embrace smarter land use (more with less) and the depletion of developable land, there is a future need for Urban Planning and Development skills. The compliment of having a strong engineering background is a great companion to this type of career track. As you have already learned, engineering requires a strong basis in the rational and the details of how something is designed – lots of left brain skills. A career in Urban Planning / Development requires the use of more of your right brain skills – possibilities, present and future focused, spatial abilities. So someone with the abilities to combine effectively both of these skills can be very useful in this field.

Deciding on a field for a master's degree is an important decision. I can tell you that a Masters in Business would be a very good compliment.

Q: I am in my third year of college. I will be receiving my associate's degree and then I will be transferring to Lamar University to pursue my bachelors in Chemical Engineering. I have always been good in Math and fair in science, and now that I'm going to a 4 year school I'm starting to feel a little discouraged about facing other students who could possibly be 10 times smarter than I am. I just want to know if the "average" student can become a chemical engineer. I looked at the degree plan and

I have to take plant design, process control, CAMS, etc., and I have never heard of any of these classes! I was told once by a chemical engineer that once I get past Calculus II things won't be so difficult for me, and I am proud to say that I will be done with Calc II in May...Yay!!! So I would like to get some feedback from other chemical engineers about some of their experiences while in school. I just want to know that I can do this even though I was not class valedictorian.

A: I completely understand your concerns as a lot of students are also intimidated by the curriculum outline for Chemical Engineering programs. However, trust me, you are anything but average. There will always be people who have talents that differ from yours...that doesn't make them "smarter" or better equipped in pursuing a chemical engineering degree. The greatest difficulty isn't the perplexing math questions or physical chemistry and thermodynamics problems, the greatest difficulty is having the self-confidence to challenge oneself to continue. Most students, like you, have worked hard and achieved high academic accolades as reflected in their grades...it's not about grades! You are going to embark on a very challenging academic journey that will challenge you mentally. There will be concepts you will grasp easily and others that you may never understand. As long as you do your best, challenge your mind, and focus on the learning rather than the grades, and strive to pursue your own standards of achievement (regardless of the grades, behaviors, or specific talents of others that differ from you) you will succeed...I promise!

Q: I am a junior in high school and beginning to look at colleges and possible careers. I am currently taking biology, pre-calculus and honors physics. I took honors chemistry as a sophomore. My grades are excellent and I like math and science. But where do I go from here? I have no interest in the medical field. One of the things I've read about engineering is that an interest in taking things apart to figure out how they work is an indicator that a person might have an aptitude for engineering. Unfortunately, this does not describe me! I spend all my free time taking dance classes! My parents have thankfully saved for my college education. I want to make the most of it. What type of engineering would be a good fit for someone like me?

A: It definitely sounds, by what you are describing that engineering is something you should explore. The fact that you like math and are good in science is a good indicator that engineering is for you. The type of engineering good for you is something you have to explore based on your likes for the gadgets and things around you. For example, are you good with computers or do they even interest you? Do you like electronics and knowing how they work? If you answer yes to these two questions then you should think of either computer or electrical engineering or even computer science. Sometimes even If you are not curious of how things work, or if you don't want to take things apart, that doesn't determine whether you will be good at engineering. There are lots of engineering fields which don't require you to take things apart.

If you like bridges and architecture and how things are built or how you can make them stronger, then you should consider architectural or civil engineering. If you like Chemistry and how to mix components, or are interested in how make-up is made, then you should consider chemical or industrial engineering. If you like biology there are lots of options for you in the engineering field, like bio-engineering and other fields related to this area. The best thing for you to do at this point is to do some research online about the different engineering fields available out there and you will see there are so many options for you to consider.

I do advise you to look into engineering as a career. It is a rewarding career and also an investment in your future.

Q: I am a 2nd year student at El Camino College and my major is Computer Engineering. I sometimes get really discouraged when most people are progressing along in their majors faster then I am. I mainly think I should just change my major and finish fast. Have you ever had these thoughts and if so how did you solve the problem?

A: Congratulations on reaching your second year in Computer Engineering! Although it is a tough, time-consuming major, when you're done, you'll have a wonderful sense of accomplishment. You will also have a marketable skill that can be applied to many real-world problems, and many companies will be interested in hiring you.

Whenever you feel discouraged, just remembered your goal. Your future goal allows you to feel good in the present about all the hard work you put into your major. I want to share this quote with you:

"If a thing is worth doing, it is worth doing well. If it is worth having, it is worth waiting for. If it is worth attaining, it is worth fighting for. If it is worth experiencing, it is worth putting aside time for."

Also, remember to learn the basics before proceeding to more advanced topics, and you will be greatly rewarded. I wish you all the best in your studies!

Q: After getting my BS in Physics I decided to spend a few years traveling and seeing the world before I settled down to a graduate program. After three years I found myself teaching English in Japan, and wondering how to get back in the game, and where. For my masters degree I want to move into engineering so I can apply my love of science to help people. But I'm not sure what field is best for me. How do you decide what field to choose? Next I have to consider location. I'd be happy to stay on in Japan and get my degree here. I also could imagine continuing my education in Europe or back home in the States. What accreditation concerns should I think about when attending an engineering school abroad? Will a Japanese or British degree be frowned on by employers or other schools in the US?

A: The questions that you ask can only be answered by you. Choosing a field is very personal. I don't know what nationality you are or what your background is, so I do not know if career tools are available to you, but in the States, there are many ways to help sort out what you should try first.

You may have heard of the Meyers Briggs test. This helps you identify your character traits and how you see the world. If you are interested, I am sure you can pick up a book on it. There are other tools that help identify what you are good at, versus what you like to do. They are not always the same. Interestingly, your scores change over time as you grow and develop new preferences.

As for where to get a degree, as you rightly point out, that is dependent on a lot of things such as money, affinity for the country and others. You have to follow your instincts on that. Note that there is a global shortage of competent engineers and scientists. The key is to make sure that you get a good education, regardless of where, and that you learn critical thinking. It would also be helpful if you have strong communication skills.

The bottom line is to look inside, versus outside. You will get 100 different opinions from 100 different people, but only you know your self and what you like and dislike. Use that as your guide on where to go next. And if you don't like it, you can always change.

Q: What does it take to become a biomedical engineer? What colleges are there for biomedical engineering? How long will it take to become one? Where are the best places to work when you get out of college?

A: A good biomedical engineer should be interested in solving biological or medical problems. Some people think of new prosthesis devices or therapeutic delivery agents as examples, but you should not limit yourself to what you have seen. You are only limited by your imagination.

There are several skills that you will need to become a true engineer with the ability to solve those questions. You will need to use a lot of math, and understand the fundamentals of physics and chemistry very well. Those subjects teach you design principles an engineer needs. And of course, you need to understand the biology of the problem you are trying to solve.

Some well-renowned biomedical engineering programs in the country include these: Johns Hopkins, UCSD, Georgia Tech, MIT, UW (Seattle).

After 4 years of college training, you would have many job options as a biomedical engineer. One or two more years of masters training might start you at a higher salary. These job functions can be engineers in a medical device company or research associates in biotechnology or pharmaceutical companies. You may work in research and development, manufacturing, or quality assurance departments. Some people prefer to go to medical school or graduate school after college. Those will take another five years before you become a medical doctor, or a Ph.D. where you can become a professor or work as a senior engineers/research scientists in the companies mentioned above.

Q: How is engineering used to prevent viruses, fight viruses, or even create viruses that you may get on your computer?

A: Anti virus software companies have engineers who develop the virus scanning software. I am sure the teams of engineers vary in their specific engineering background but primarily consist of software engineers. As far as creating viruses there may be certain situations where software engineers may want to create a virus to simulate how secure their product is or how well their scanner may detect a specific type of virus.

Q: I am in school and am pursing my Associates in Architectural Engineering and have no experience in the field. I have worked in Accounting and Financial positions for the most part but would like to find a job in a related field while I finish school. Are there any entry level jobs in these fields? How would you suggest going about this? Also, although I am a very good student thus far, I am one of very few women in this field at my school; do you think this will have a major effect on my job opportunities after graduating? Thank you.

A: I think it is wonderful that you are pursuing a career in Architectural Engineering. I also think it is ideal that you have both an accounting and finance background, a combination that is rare to find with someone interested in engineering! I sincerely believe that engineers of the 21st Century should be trained with more than technical subjects including project management, leadership and public policy, among others.

I would encourage you to continue to complete your education in Architectural Engineering. I also would suggest you consider looking into the areas of "green design and construction" and sustainability, which are already considerations every engineer should take into account in today's environment. When you are looking at potential employers, be sure to point out that you have skills that go beyond those learned in architecture and engineering and that you understand the management aspect of the work to be performed, including the financial and accounting side of things. This will be very attractive to future employers. Regarding how you should go about this really depends on what area you wish to work in, both from an industry perspective as well as location, including whether you have an interest in international work. However, a good source of information is Engineering News Record which has information on most all design and construction companies in the world and the industries and areas of the world in which they work. Their website is www.ENR.com.

Q: I will be getting my bachelor's degree in geography. I'm really interested in becoming a civil engineer, but it is too late to change majors. Is it too late to become an engineer? If not, what path do you recommend I take? Go back and get another bachelor's degree in engineering? Will it take four years again or will some schools accept my previous credits? Or go to a community college for two years to take math and science, and then transfer to a four year school? Or will some graduate schools accept me without a BS engineering degree? I'm just anxious because though I'm good at math, science isn't my strong area, especially chemistry. And I feel like I'm too late. I will be older than everyone else in my class. But I really want to help improve the environment and the world around me. What do you think?

A: It is never too late to become an engineer! I encourage you to pursue your interest and explore what options are available to you. Your strong skills in math will certainly support your pursuit of an engineering degree. I wouldn't let chemistry hold you back.

First, consider contacting the engineering department at your current school to review how many of your undergraduate classes would count towards a BS in Civil Engineering. Most likely, you have taken a good portion of the first/second year requirements already and another 4 years would not be necessary. Secondly, graduate programs in engineering often have students from other disciplines. However, given your background in Geography, there may be some undergraduate classes that would be required in order to transition into a Civil Engineering graduate program.

It is hard to recommend which path to take, without knowing more about your specific situation. I can offer a few pieces of advice during your decision making: 1) Make a Pro/Con list, actually write them down. It is amazing how easy some decisions are when you see it on paper! 2) Choose an engineering program with a good intern/co-op program to gain some "real-world" experience while you finish your degree. 3) If financial support is a major decision-maker, investigate potential scholarships through organizations (e.g. Society of Women Engineers).

Conclusion

This chapter gave you a good idea of how to find the right type of engineering for you. It provided you with practical advice on volunteering, summer camps and work study programs. Your personality traits and abilities were examined to provide you with insights into what is needed in each specialty.

Contractors and consultants were defined and you now have a better idea as to where they fit into the engineering profession and if you are interested in this type of work.

Your strengths and skills were examined and evaluated to help you find the right engineering field for you. Engineering is an important and learned profession. You now know that Engineers must perform under a standard of professional behavior that requires adherence to the highest principles of ethical conduct and a summery of the Code of Ethics for Engineers was given in this chapter.

EDUCATION AND LICENSING

Education

You will need a BS degree for any type of engineering you decide to go into. A master's degree or even a PhD may be necessary to obtain some positions in research, education, or administration in most engineering fields. Most engineering degrees are granted in electrical, electronics, mechanical or civil engineering. College graduates with a degree in the physical sciences or mathematics occasionally may qualify for some engineering jobs. Chemical and Material engineering degrees are probable the most difficult of all the engineering degree.

However, engineers trained in one branch may work in related branches. Many aerospace engineers have their degrees in mechanical engineering. This flexibility allows employers to meet staffing needs in new technologies and specialties in which engineers may be in short supply. It also allows you to shift fields with better employment prospects.

In undergraduate school, expect to take 4 years of higher mathematics and 4 years of physics plus a year of chemistry. In the first two years you'll take the majority of your liberal arts classes while during the second two years, you will take the majority of classes in your degree specialty. At some schools a five or six year program combines classroom study with practical experience working for an engineering firm. Many colleges also offer dual major programs where you can graduate with degrees in two related fields.

Most engineering programs concentrate in the engineering specialty along with courses in both mathematics and science. Most programs include a design courses sometimes accompanied by a computer or laboratory work.

A graduate degree is a prerequisite for becoming a university professor or researcher. Also, more and more large corporations only hire new grads with Master's degrees. New graduates with MS degrees will always command higher starting salaries. A substantial number of engineering graduates combine their engineering degree with a master's degree in business administration (MBA). The engineering/MBA combination is a powerful one for engineers who expect to move into management, marketing, sales or plan on starting their own business. Engineers whose work may affect the life, health or safety of the public must be registered according to regulation in all 50 states and the District of Columbia. You must have received a degree from an accredited engineering program and have four years of experience. You must also pass a written examination. Many engineers also become certified. The doctorate degree (PhD) is obligatory for engineers who expect to teach at the university level.

Picking a College

When you make plans to attend college and study engineering you should consider several factors: the type of engineering program, theoretical verses practical, , the quality of the school, the quality of the instruction, the employment possibilities, the cost, the availability of finical aid and your personal preferences for living and learning conditions.

In the United States there are more than 300 colleges and universities where engineering programs have been approved by the Accreditation

Board for Engineering and Technology (ABET). It is wise and will help you with licensing if you attend a ABET accredited engineering program. Make sure all the departments or programs within the college are ABET accredited. Although admissions requirements vary, most require a solid background in mathematics and science as well as good SAT scores, especially in Mathematics. ABET accreditation is based on an examination of an engineering program's student achievement, program improvement over time, faculty, curricular content, facilities, and institutional commitment.

Although most educational institutions offer programs in the major branches of engineering only a few offer programs in the smaller specialties. Also programs of the same title may vary in content. For example, some programs emphasize industrial practices preparing student for a job in industry, whereas others are more theoretical and are designed to prepare students for graduate work. Therefore you should investigate the curriculum and check accreditations carefully before selecting a college.

Undergraduate and Graduate Programs

Admission requirements for undergraduate engineering schools include a solid background in mathematics (algebra, trigonometry and geometry) and science (biology, chemistry and physics) and courses in English, social studies, humanities and computer and information technology. If these are not taken during high school, you may be required to take remedial studies to make up the deficiency. Bachelor's degree programs in engineering typically are designed to last 4 years and usually require 132 college credits. , but many students find that it takes up to 5 years to complete their studies. Some colleges offer master's degree programs. Some 5 year or even 6 year cooperative plans combine classroom study and practical work. Most college admission departments will look for higher mathematics SAT scores rather than English SAT scores. However, both Math and English scores should be higher than average.

Like many graduate programs, colleges of engineering require the Graduate Record Examination (GRE). You should take the GRE during your senior year of undergraduate school even if you do not think that you will be going on to graduate school. The scores are good for five years and having taken the exam keeps all of your options for the future open. It can be intimidating to take the exam when you have been out of school for several years.

An engineering education will teach you quantitative reasoning, problem solving and design. Your overall college education will teach you critical thinking, writing, and speaking, all of which can successfully be applied to a number of different jobs in the field of engineering. But it still remains up to you to choose a job and to learn how to articulate the benefits of your education in a way that a potential employer will appreciate.

Alternate Programs

Most high school graduates going to college plan to attend for four years, take summers off, and graduate. However, there are a variety of alternatives. One of the more important of these for engineering students is the cooperative program. This is a method of paying for school while you are attending and getting training that might be directly related to your career. Usually each term is alternated with a term of working with a local employer. The work can range from technician's tasks to being a junior engineer who assists working engineers. Coops can be a good idea in this economy.

Cooperative or work study programs take longer but they offer financial advantages. Note that BSE degrees are not as strong as BS degrees when you get into the business world so if you can afford to go to a 4 year undergraduate school, you will be much better off. BSE degrees are less theoretical in nature and you will take a smaller number of classes in your specialty. Many BSE graduates are not well prepared for original design work and will find themselves in Test or Manufacturing Engineering.

Another type of program that some students use is a 3-2 program leading to a master's degree. This often exists at schools that do not have a fully accredited engineering curriculum. The school accepts students who take all the preparatory courses for an engineering degree during the first three years. The student then attends another school where the necessary engineering courses are taken.

Internships that enables you to get a security clearance and a lot of engineering experience are going to help you in certain engineering fields such as electrical or aerospace.

Traditional and Specialties of Engineering

Engineering is a very diverse and challenging field of study. With more then 25 major branches of engineering and 100 specialties, there is something for everyone who pursues the field. Here are a few major branches of engineering and the specialties that are within them that you can get your degree in.

1. *Biomedical Engineering* is a major branch which then gets extended to biomechanical, bioelectrical, biochemical, rehabilitation, clinical and genetic engineering.
2. *Chemical Engineering* is the traditional; biochemical, food, and materials engineering is the extension of it.
3. *Civil Engineering* has architectural, environmental, structural, and transportation engineering as its extensions.
4. *Electrical Engineering* specialties are computer, software, hardware and optical engineering.
5. *Mechanical Engineering* is the traditional field and its specialties are industrial, agricultural, aeronautical, aerospace, automotive, heating, ventilating, refrigeration, air conditioning and manufacturing engineering.

Engineering Licenses

All 50 states and the District of Columbia require licensure for engineers who offer their services directly to the public. Engineers who are licensed are called Professional Engineers (PE). This licensure generally requires a degree from an ABET-accredited engineering program, passing the Fundamentals of Engineering (FE) exam formerly called the Engineering In Training (EIT) exam, four years of relevant work experience, and successful completion of a State Principal of Practice of Engineering (PE) examination.

Recent graduates can start the licensing process by taking the examination in two stages. The initial Fundamentals of Engineering (FE) examination can be taken upon graduation. Engineers who pass this examination commonly are called Engineers in Training (EIT) or Engineer Interns (EI). After acquiring suitable work experience, EITs can take the second examination, the Principles and Practice of Engineering exam. Several States have imposed mandatory continuing education requirements for re- licensing. Most States recognize licensure from other States provided that the manner in which the initial license was obtained meets or exceeds their requirements. Many civil, electrical, mechanical and chemical engineers are licensed PEs.

An engineering license is not necessary when practicing an engineering discipline in private industry, however, most companies will not grant engineering titles to employees who have not graduated with an engineering degree.

Tips for Meeting the Licensing Requirements & for taking the Exams

The first thing is to make sure you college or university has an ABET accredited engineering program. You need to be careful because some departments or programs within a college or university can be ABET accredited while others are not. The next step in the licensing process will be passing the Fundamentals of Engineering (FE) exam. It is administered every Fall and Spring by state engineering registration boards.

Most civil engineers go on to study and qualify for a professional engineer (PE) license. It is required before you can work on projects affecting property, health or life. Because many engineering jobs are found in government specialties it is a good idea for you to get this license. Registration guidelines are different for each state and they involve educational as well as practical experience.

You may benefit if you take the FE exam before graduation. You will have accumulated a large amount of knowledge during your schooling and two or three years after graduation it can be extremely difficult to remember concepts and facts necessary to pass the FE exam. If you are currently a student in an accredited engineering program chances are that your school provides a review course or other assistance in preparing for the FE exam.

If you are already out of school you will have to arrange your own study program. Some colleges and universities offer extension courses designed for FE candidates and for PE candidates in some fields.

State chapters of the National Society of Professional Engineers (NSPE) and some technical societies give refresher courses. In some areas, organizations such as the Professional Engineering Institute and local engineering clubs offer review courses. Particularly, if you have been out of college for some time or if you are not a graduate of an accredited engineering program, a review or refresher course can be extremely helpful. There are also a number of books written specifically to assist in preparing for these exams.

It is necessary to apply to take the exam well in advance. You should contact the engineering registration board in the state or sates where you plan to become licensed prior to starting your senior year of college.

The FE and PE Exams

The FE exam covers a broad spectrum of engineering knowledge which means you must review basic science, mathematics, as well as many different fields of engineering. It consists of two four hour sessions. The morning exam tests comprehension and knowledge as well as evaluation, analysis, and application. The afternoon exam is composed of problem sets from seven subject areas, statics, dynamics, mechanics of materials, fluid mechanics, electrical theory and economic analysis.

It is an open book test; however states do vary in the amount of material that you are allowed to bring into the exam. This is important information to ask when you contact the state board of registration regarding test dates and deadlines. After you pass the FE exam and obtain the necessary work experience, you will be ready for the PE exam. This exam will test your in-depth knowledge of a specific field of engineering.

There are important differences between the scoring methods used with the FE exam and those used with the PE exam. Some of these differences are due to the different formats of the two exams. For instance, partial credit is impossible with the multiple choice questions, while partial credit can be given on the essay problems in the PE exam. Also, only one answer is correct for each multiple choice item on the PE exam, whereas problems on the PE exam like problems in real life engineering may have more than one correct solution.

Statistics from recent years indicate that roughly one half to two thirds of all candidates who take the PE exam pass it, depending on the discipline. It is much more difficult with the PE exam than with the FE exam to make comparisons among candidates from different specialty areas. In some areas there are many candidates and in others there are a few. However graduates of ABET accredited engineering programs do better on the average than candidates who come from other backgrounds.

There are some situations where you can be exempted from the PE exam. You may register as a professional engineer without taking the standard set of tests if you have already registered through the normal examination route in another state, if you qualify under a grandfather clause or if you may be considered eminent or if you can claim a long established practice. If You Fail the PE

Don't get too upset, take a review course and take it again. In most states you can take the next exam that is offered. However, if you fail more than once you may be required to wait a specific amount of time before trying again.

License Requirements

Licensing is not required for most engineering jobs in corporations except when you will be working on public projects. However it is recommended to enhance your credentials and make you appealing to more job opportunities. License requirements for engineers usually include a degree from an American Board for Engineering and Technology accredited engineering program, four years of work experience and successful completion of a state exam. There are many different types of engineering and ways to become licensed. Here are a few examples.

Few schools offer an undergraduate degree in environmental engineering. Another way to become an environmental engineer is to earn a civil, mechanical, industrial or other traditional engineering degree with an environmental focus. You must pass an Engineer In Training (EIT) exam covering the fundamentals of science and engineering. A few years after you've started your career, you also must pass an exam covering engineering practices. Additional certifications are voluntary.

Not all computer engineers are certified. The deciding factor seems to be if certification is required by the employer. Many companies offer tuition reimbursement or incentives to those who wish to earn certification. Certification is available through the Institute for Certification of Computing Professionals and the Associate Computing Professionals (ACP) and the Certified Computer Professionals (CCP). Certification is considered a sign of industry knowledge. An option if you are interested in software engineering is to pursue commercial certification. These programs are usually run by computer and networking companies that wish to train professionals in working with their products. Some computer certifications are available through the companies offering the products such as Microsoft, Oracle, SAP, Cisco, etc. These are recognized as demonstrated skill in managing these products.

Licensing as an industrial engineer is recommended since an increasing number of employers require it. It will also be in your favor to have it when applying for positions in the public sector. Licensing requirements vary from state to state but in general they require you to have graduated from a accredited school, have four years work experience and have passed the eight hour Fundamental of Engineering Exam (FE). At that point you will be an Engineer in Training (EIT). Once you have fulfilled all the licensing requirements you receive the designation of Professional Engineer (PE)

For Packaging Engineers, the Institute of Packaging Professionals, a professional society offers two levels of certification: Certified Professional in Training (CPIT) and Certified Packaging Professional (CPP). The CPIT is available to college students, recent graduates and professionals who have less than six years of experience in the field. Requirements for this certification include passing a multiple choice test and an essay test. The CPP can be earned by those with at least six years of experience in the field. In addition to the experience requirement, candidates must fulfill two other qualifications from the following: presenting a resume of activities, writing a professional paper or holding a patent. Although certification is not required, it is a good idea to obtain it to show that you have mastered specified requirements and have reached a certain level of expertise.

Most plastics companies do not require a bachelor's degree in plastic engineering. Companies that design proprietary parts usually require a bachelor's or advance degree in mechanical engineering or chemistry. The field of plastics engineering overall is still a field in which people with proper experience are scare and experience is a key factor in qualifying you for an engineering position. To obtain a bachelor's degree in plastics engineering you should contact the Society of Plastics Industry (SPI) or the society of Plastic Engineers (SPE) for information about four year programs. National certification is not required. Both SPE and SPI have established voluntary certification programs.

Although there are no licensing or certification requirements designed specifically for quality control engineers some need to meet special requirements that apply only within their industry. Most Quality Control Engineers come out of the industries from which they are employed. Many quality control engineers pursue voluntary certification from professional organizations to indicate that they have achieved a certain level of expertise. The American Society for Quality offers certifications including Quality Engineer Certification (CQE). Requirements include having a certain amount of work experience, having proof of professionalism such as being a licensed Professional Engineer and passing a written examination. Many employers value this certification and take it into consideration when making new hires or giving promotions.

The Institute of Transportation Engineers (ITE) offers certification as a Professional Traffic Operations Engineer (PTOE). To become certified you must have at least four years of professional practice in traffic operations engineering, hold a valid license to practice civil, mechanical, electrical, or general professional engineering and pass an examination.

The need for continuing education is a common denominator in the engineering professions.

Examples of Education Needed for Certain Engineering

Here is an example of the education you would need if you wanted to be a Systems Engineer.

College degree programs in systems engineering are diverse, giving you the chance to interact with a variety of disciplines. You can choose to specialize after the first couple years of study, but this is usually not required. Also, if you find an engineering discipline that suits you better than systems engineering, it is relatively easy to transfer credits over to a different program or to switch majors.

Systems engineering degree programs include the Bachelor of Science in Systems Engineering, the Master of Science in Systems Engineering, and the Doctor of Philosophy in Engineering, as well as specific certificate programs covering a broad range of topics. Each of these degrees can lead to rewarding careers as entry-level engineers, project managers, or teachers. The flexibility of the systems engineering degree means that careers are available to graduates in almost every industry.

Many professionals already within the field of systems engineering don't have the time or the need for a full degree program. For this reason, many certificate programs are available for people wishing to concentrate on a single area of systems engineering. Many of these certificate programs are available online.

Bachelor of Science in Systems Engineering

The first two years of most BS degree programs in systems engineering provide a basic overview of the field of engineering in general, with specializations available in the third or fourth years. This overview includes the study of technology and science, design engineering, and business.

The technology and science sections usually cover core mathematics and computer science topics, as well as physics and chemistry in the context of engineering. The design portion usually covers basic building techniques, and designing with different qualities in mind, such as function versus durability. Studying business provides valuable insight into the larger context in which engineering exists, and teaches students how to thrive as engineers in the business world.

During their first two years, systems engineering students are often able to take one or two elective courses in something that interests them specifically, from history to philosophy to auto mechanics. The third and fourth years of study are often supplemented by a research project, a work-study project, or a term abroad, depending on the program. You'll have the opportunity to begin gaining hands-on experience in the field before graduating, thereby making the post-graduate job hunt much easier. These projects can range from working as an engineering assistant or research assistant to traveling abroad and working on simulated systems using a foreign platform.

Furthermore, you should be able to customize your program to a higher degree by selecting appropriate electives or picking up a minor in another discipline of engineering, such as business engineering or automotive engineering. You can now begin to gear your degree toward the industry you want to work in.

Online BS degrees are a natural choice for non-traditional students, students who wish to work full-time, and students who can't leave home for one reason or another. These e-learning degrees are available from fully accredited online universities across the country.

Master of Science in Systems Engineering

Most master's degree programs in systems engineering are less about taking core classes, and more about developing a project that acts as the equivalent to a master's thesis. These projects are intended to simulate the process of working as part of a team in a particular industry, such as robotics, biomedical engineering, or electronics.

The projects are usually based on solving a specific systems problem or set of problems. Graduation is dependent upon the successful solution of the given problem or problems, and there is usually a presentation involved, as well as supervision, observation, and review during the project itself.

Pursuing a graduate degree in systems engineering is an excellent way to get hands-on experience prior to entering the work force. While most students will have some general hands-on experience due to lab work at the undergraduate level, the master's level project is a way to gauge how your problem solving capabilities and understanding of systems measures up.

Graduate programs, particularly online ones, often offer more flexibility than traditional undergraduate programs, allowing you to maintain a job or spend time with your family. With fewer and more flexible classes to attend, graduate students can make their own schedules. This means that they can devote more time to job and family, while doing the research project at their convenience.

Doctorate Degree Programs in Systems Engineering

If you wish teach at the college level or to become a research professional in your field, you may further your studies by entering a PhD program in systems engineering. The Doctor of Philosophy in Systems Engineering is a terminal degree, making you eligible to teach at the university level.

Most PhD programs do not have many required classes, but are instead based on a single large project known as a dissertation. In order to be eligible for the PhD, you must pass an extensive exam that covers all of the material you learned in undergraduate and graduate school. If you pass, you may submit a dissertation proposal.

If the dissertation proposal is accepted, you must complete the dissertation and be prepared to successfully defend it. Doctoral candidates are usually required to complete at a least one-year residency. During this time they must find funding (usually external funding) for their dissertation research.

The dissertation can be on any topic you wish, although your choices are likely to be guided by the research specialties of your faculty (something to keep in mind when choosing a PhD program). Since the range of available topics is very broad, doctoral candidates can choose a subject important to them, in a field in which they believe they can make a difference. It is important to pick a topic in which you excel, as dissertations are lengthy undertakings. Dissertations are not just designed to earn a degree; they are valid forms of research conducted by experts at the top of their profession.

What can you do with a Major in Systems Engineering?

Career options for aspiring systems engineers

As production systems become more complicated and more businesses change their methods to accommodate the global marketplace, the need for systems engineers has grown tremendously. Since systems engineering is a multidisciplinary field, jobs exist in nearly every industry. This gives you the chance to explore many types of businesses in order to discover which company is the best suited to your interests and special talents.

As production moves overseas and the parts involved in assembling products come from all over the globe, the opportunity for travel is great in the field of systems engineering. While travel can often be exhausting, it also helps the engineer broaden their horizons and achieve a better understanding of how to integrate things into a cohesive system.

Some examples of the roles systems engineers might play in a variety of industries include:
- Petroleum systems engineers seek out oil and gas deposits, then figure out how to extract, store, and transport the materials in as safe a manner as possible. Petroleum systems engineers are engaged in every activity an oil company is engaged in, overseeing the drilling, processing, and equipment maintenance. These specialists are constantly on the lookout for ways of making the process more efficient, cleaner, or safer.

- With a focus on solving problems in industrial systems, industrial systems engineers find ways of streamlining a process, making it more cost-effective. They also focus on making the production floor a safer place to work. They must understand the mechanics, physics, and chemistry behind all kinds of machines, while designing solutions for problems. In addition, these specialists must communicate their recommendations to everyone from the Board of Directors to the other engineers on the project.

- As developed nations finally begin to address their impacts on the environment, the demand grows for environmental systems engineers, who have a strong feeling of social responsibility and an understanding and concern for nature. Industries such as waste disposal, wastewater treatment, water purification, and emissions utilize environmental systems engineers. These jobs are often in the public utilities sector. These specialists are very well respected because of the social agendas they promote.

• Getting computers using different platforms to work together and communicate can be a frustrating task. This is where software systems engineers come in. They create software that will not only work on as many operating systems as necessary, but will also communicate effectively with other operating systems in the industry. Often, these specialists must bridge the communication gaps between suppliers and customers. These engineers also create software for specific industries in order to regulate or control certain processes, such as an assembly line or quality control facility.

• Electronic systems engineers work in all sorts of electronics industries, including telecommunications, microelectronics, and robotics. They are involved in everything, including the design, implementation, operation and maintenance of new systems. These systems range from the automated production of an item to the regulation of company's information network.

Certification and Licensure

The law does not specifically require systems engineers to be licensed, but it does make looking for a job much easier. If you are working for the federal government, then you need to be a licensed engineer, and most states have regulations concerning what sorts of jobs only licensed engineers can take. Therefore, it is best to get licensed in your state, if possible. Licensure regulations also vary from state to state, and students can learn their state's exact requirements through their school's career counseling program.

In general, licensure usually consists of passing a state exam, but most states require applicants to have a few years of work experience before sitting for the exam. If you're fresh out of college, you can register for pre-licensure certification. Once you have gained the required work experience, you can apply for actual licensure.

There are also a number of trade and professional associations for engineers that can help those in the field to network and to share ideas. These organizations offer seminars, luncheons, and job fairs, so engineers can explore their options and keep current on what is happening in their profession.

Career Education in Environmental Engineering

In planning for your career as an environmental engineer the majority of college degree programs in the field of environmental engineering exist at the bachelor's and master's level (particularly online degree programs). Associate's degrees and undergraduate certificate programs are less common.

Bachelor's Degrees in Environmental Engineering

A bachelor's degree in environmental engineering is required for gaining employment as an environmental engineer. The bachelor's degree emphasizes math and science courses as well as classes specific to the environmental engineering field. Examples of these courses are air pollution engineering, environmental risk assessment, and principles of environmental engineering.

At some universities, environmental engineering is a supplementary program to a bachelor's degree in civil, chemical, or mechanical engineering. In these cases, students earn bachelor's degrees in another branch of engineering with a minor in environmental engineering. A bachelor's degree in environmental engineering typically takes five years to complete, though some students may be able to complete it in four.

Some colleges and universities offer advanced online certificates in environmental engineering studies. This certificate typically focuses on one aspect of environmental engineering and offers four to five courses in that area. It is intended for bachelor's degree holders who wish to continue their education beyond the undergraduate level. It is not considered a graduate degree.

Master's Degrees in Environmental Engineering

To obtain your master's degree in environmental engineering, you must first have a bachelor's degree in engineering or a related field, such as one of the sciences. Courses required vary depending on the college or university you attend, but courses in management, protection of resources, pollution control, and water quality are common. A master's degree typically takes two years to finish; online degree options have increased in number and popularity in the last few years as working professionals study without leaving their jobs.

What can you do with a College Major in Environmental Engineering?

Career options for aspiring environmental engineers

- An engineering technician is an assistant to an engineer or scientist. In the environmental engineering field, he is often responsible for assisting in research, collecting data, maintaining equipment, and assisting in the planning and execution of projects. Engineering technicians usually have an associate's degree in engineering technology and are not required to have a license. An engineering technician may be required to work in a hazardous environment, such as dealing with nuclear waste removal or waste treatment.

- Environmental engineers resolve and help prevent environmental problems. They work in many areas, including air pollution control, industrial hygiene, toxic materials control, and land management. The duties of an environmental engineer range from planning and designing an effective waste treatment plant to studying the effects of acid rain on a particular area. An environmental engineer is sometimes required to work outdoors, though most of their work is done in a laboratory or office setting. Career opportunities for environmental engineers exist in consulting, research, corporate, and government positions. At minimum, environmental engineers must possess a bachelor's degree. Master's degrees are strongly encouraged, but not required. Environmental engineers offering their services directly to the public must be licensed.

• Engineering managers supervise engineers and support staff. They typically begin their careers as engineers and advance to the managerial level. Engineering managers are responsible for administrative work in addition to supervising staff and engineering projects; these tasks often involve budgeting, creation of policies and procedures, and the hiring and training of staff members. In the field of environmental engineering, most managers hold office jobs, though some may work in a laboratory setting. Engineering managers often receive benefits such as stock options and bonuses. In addition to their engineering degrees, they typically have some business or management training.

• Environmental sales engineers are responsible for selling equipment and/or services related to the environmental engineering field. For example, a sales engineer in the environmental engineering field may be responsible for the sale of air pollution control products to factories. Sales engineers bring their education and experience to the position, enabling them to speak with their customers clearly and accurately about the products they sell. In addition to sales, they often assist the Marketing department with the design and modification of their products based on customer feedback. Careers in sales can be stressful because, in most cases, quarterly job performance directly impacts job security and earnings. Sales engineers are often required to travel, sometimes long distances. Those engineers required to travel are often compensated with company cars and accumulation of frequent flyer miles.

Salary Expectations for Environmental Engineers
According to the U.S. Bureau of Labor Statistics, environmental engineers earned a median income of $69,940 in 2006. Engineering technicians' salaries vary widely by industry but the median hovers between $40,000 and $50,000 per year.

Certification and Licensure
Environmental engineers are strongly encouraged to become Licensed Professional Engineers. Requirements for this licensure vary from state to state, but typically involve:
• graduation from an accredited engineering program,
• work experience, and
• passing the state's Professional Engineering examination

Conclusion

You now know you will need at least a BS degree for any type of engineering and a master's degree or even a PhD for some positions. How to pick the right college for you was discussed and alternative programs were evaluated. The traditional engineering majors and their specialties were outlined and you now have a better understanding of what you should major in and why.

Licenses for engineers were defined and you know the different types of licenses along with when, why and how to get them. Also, you understand when you need to be licensed and when it is not necessary. The Fundamental of Engineering, Professional Engineering and other licenses were discussed. The exams for these licenses were described in detail and helpful hints on passing these exams were given.

Different types of engineering and their degree requirements and licensing were discussed. Engineering sectors such as environmental, computer, industrial, packaging, plastics, quality control, and traffic operations were just a few that were described in this chapter.

You must accept that to be an engineer means to continue your education and learning throughout your career. Your value depends on your knowledge of the latest technology.

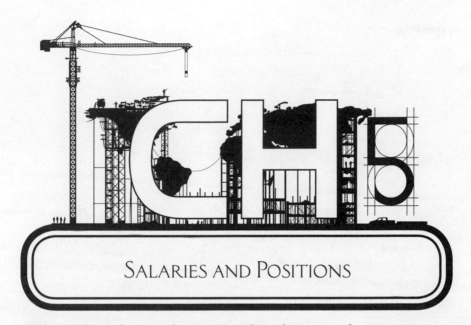

SALARIES AND POSITIONS

N ow that you have some idea what type of engineering you want to go into, you need to find out the overall growth in that engineering sector and how to obtain a position. This chapter will tell you how to get that position and what the overall industry growth will be. First, in order to get that position the main thing to do is network, network, and network. That is the key. Only 15-20 percent of all jobs are formally advertised, which means 80-85 percent of available jobs do not appear in published channels? Networking will help you become more knowledgeable about all the employment opportunities available during your job search. Everyone you know is part of your network, even if they are not engineers. They may be able to introduce you to someone they know who works in the engineering field you're targeting.

One of these people may offer you your next job, or introduce you to someone else who can. Keep your eyes and ears open, you can hear about opportunities as easily at a family gathering as at a conference. Don't be afraid to step up and explain your interest in engineering. That's the magic process of networking and it is the essential ingredient in any job seeking strategy.

Networking Rules

Networking involves some rules and etiquette. Begin by asking fifteen of your own professional, school, and personal contacts to give you names of three people they know who work in engineering. Agree on how you should introduce yourself. Ideally, your friend or colleague will offer to pave the way by calling or writing a letter of introduction. At a minimum, make sure you have your contact's permission to use his or her name whenever you call a networking referral.

Send each of your new contacts a package containing a cover letter describing your career goals, a resume and the type of engineering you are interested in. You could start with a phone call to see whether they are interested in your information or have any other suggestions. In the course of an informal telephone or face to face chat, ask if he or she knows anyone else who works for any engineering companies you are interested in. If not, ask if there is anyone else in his or her circle of business associates who might.

Recruiters

Most engineers obtain their first position through company recruiters sent to college campuses. If you find yourself face to face with a representative of the company's human resources department you will want to emphasize your interpersonal skills and willingness to learn rather than engineering lingo. If your interpersonal skills could use a little fine tuning have someone in the placement office or a friend videotape you in a mock interview.

Alternative Position Finders

You can also find employment through a summer position or work study arrangement. Cooperative experience in college lends a big hand not only because of the exposure to the actual field you're interested in but also the possibilities of after college contacts. Keeping in touch with fellow engineers leads to even more contacts. You should respond to advertisements in professional journals or newspapers. The internet now offers multiple opportunities to job seekers. Join a professional engineering organiza-tion and use their employment services. Also look into employment agencies and headhunters that deal with engineers. Interview with everyone for experience and possible connections.

General Expected Industry Growth

Depending on what type of engineering you want to go into, the demand will vary. The general overall growth of engineering is expected to grow more slowly than the average for all occupations over the 2006 to 2012 periods. Engineers tend to be concentrated in slow growing manufacturing industries, a factor which tends to hold down their employment growth.

Despite this, job opportunities in engineering are expected to be good because the number of engineering graduates should be in rough balance with the number of job opportunities over this same period. Some areas will see short term growth and still others will bloom. Not since the dot-com boom of the late 90s have job opportunities been so bright for engineering grads. The civil and construction-related engineers, however, appear to be more sought-after than their peers in electrical and software engineering. A rebounding economy and baby boomer retirements have spurred demand and boosted salary levels for engineering graduates.

Projections range from a decline in employment of mining and geological engineers, petroleum engineers and nuclear engineers to much faster than average growth among biomedical and environmental engineers.

Outsourcing

Also, many employers are increasing their use of engineering services performed in other countries, known as out sourcing. There are a large number of well trained often English speaking engineers available in many countries that are willing to work at much lower salaries than engineers in the United States. In 2004 the United States graduated roughly 70,000 undergraduate engineers, while China graduated 600.000 and India 350,000. These massive numbers of Indian and Chinese engineering graduates include not only four year degrees, but also three year training programs and diploma holders. These numbers have been compared against the annual production of accredited four year engineering degrees in the United States.

American employers are finding that engineers that U.S. colleges turn out are often more expensive than those they can hire in India. Some major companies feel when they have to look for deep technical talent, not just 10 or 20 people, especially in high technology, the places they can go and know they can hire somebody every day are India and China. Half of IBM's 190,000 engineers and technical experts now reside overseas, for instance. Of course, some engineering jobs will always remain in the U.S.

The rise of the Internet and other electronic communications systems has made it relatively easy for much of the engineering work previously done by engineers in this country to be done by engineers in other countries, a factor that will tend to hold down employment growth. How much longer can the US maintain technological edge when other nations are producing more and more engineers? Many argue that we are actively fueling this process by outsourcing American engineering jobs overseas.

Outsourcing creates a threat to certain some types of engineering. Jobs of transactional engineers are easily outsourced and are routinely being taken by relatively low paid engineers in countries like India and China. The great majority of engineers involved in outsourced professions hold a minimum of a four year degree. However, the outsourcing of high level engineering and IT professions is another story. These jobs often require specialized dynamic engineers: individuals with strong interpersonal skills, technical knowledge and the ability to communicate across borders. Foreign engineers trained by accredited universities with high language proficiencies and close proximity to their country's industrial and commercial centers are the most likely to compete wit US based engineers for offshore engineering jobs. Dynamic engineering jobs are difficult to outsource, individuals with these skills are always in demand.

Engineers will continue to play a prominent role; the real question is where these engineers will be located. The United States is producing a competitive number of engineers,. The challenge for the United States over the next decade will be to retain its role as a global pacesetter in the education of engineering.

Many job openings will come from replacement needs. Job openings will be created by engineers who transfer to management, sales, or other professional occupations. Additional openings will arise as engineers retire or leave the workforce for other reasons.

It is important for you to continue your education throughout your career because much of your value to your employer depends on your knowledge of the latest technology. If you become an electronics engineer you will find that technical knowledge can become outdated rapidly.

Even if you continue your education you are vulnerable to a layoff. If the particular technology or product in which you have specialized becomes obsolete. You must however keep current in your field. If you have not kept current you may be passed over for promotions, etc.

On the other hand these high technology areas offer the greatest challenges, the most interesting work, and the highest salaries. Therefore, the choice of engineering specialty and employer involves an assessment not only of the potential rewards but also of the risk of technological obsolescence.

Expected Growth in Traditional Engineering Fields

Some old traditional standards but still winners in the engineering profession are chemical, civil, electrical, and mechanical engineers. Majoring in these fields will open up many doors for you.

Chemical engineers will find more opportunities opening up, particularly in specialty chemicals, pharmaceuticals and plastics materials as chemical companies research and develop new chemicals and more efficient processes to increase the production of existing chemicals,. Certain corporations such as chemical companies will need engineers in research and development to work on new chemicals and more efficient processes. Areas in manufacturing that should be promising include pharmaceuticals, biotechnology and electronics.

Additional growth will come in service industries such as companies providing research and testing services. Chemical engineers work in a variety of places but most work in offices, laboratories, or plants. Traveling to other facilities can be required.

Employment for civil engineers is expected to increase more slowly than the average. Spurred by general population growth and an increase emphasis on infrastructure and security, more civil engineers will be needed to design and construct safer and higher capacity transportation, water supply and pollution control systems.

Employment in civil engineering will come from the need to maintain and repair public works, such as highways, bridges and water systems. In addition, as the population grows so does the need for more transportation and pollution control. Firms providing management consulting and computer services may also be sources of jobs in civil engineering. However, employment is affected by several factors such as decisions made by the government to spend further on renewing and adding to the country's basic infrastructure and health of the economy in general. You can find a position in civil engineering almost anywhere.

Many civil engineers work regular weeks often in or near major industrial and commercial areas. You sometimes will be assigned to work in remote areas or foreign countries. Offices, labs, factories, and actual work sites are typical environments. About one third of civil engineers work for various levels of government usually involving large public works projects.

You will find the field of electrical engineering fluctuates with the changes in the economy. If there is a growing interest in defense related fields, upgrading aircraft and weapons systems, use of electronics components in automobiles and increases in computer and telecommunication then opportunities for electronics engineers will increase. If you become an electrical engineer you may work in a laboratory if you are involved in research. Or you may work in an office and may spend part of your time in production facilities. You may also encounter extensive travel if you get involve in field service and sales. Government security clearance is a big plus.

Although overall employment in manufacturing is expected to decline, mechanical engineers will be needed to meet the demand for more efficient industrial machinery and machine tools. Employment for mechanical engineers in business and engineering services firms is expected to grow faster than the average for other industries. Also, increases in defense spending may create improved employment opportunities for engineers in the federal government.

Employment for mechanical engineers is projected to grow more slowly than the average for all engineering occupations. But employment of mechanical engineers in manufacturing should increase more rapidly as the demand for improved machinery and machine tools grow and as industrial machinery and processes become increasingly complex.

Most mechanical engineers work indoors in offices, research laboratories or production departments of factories and shops. Depending on the job, a significant amount of work time can be spent on a factory floor, at construction sites, or at other field operations sites.

Expected Growth in Specific Engineering Fields

Employment is projected to grow more slowly in aerospace engineers than in industries generally. Earnings are substantially higher, on average, than in most other manufacturing industries Foreign competition and slowdowns in air travel will limit the number of new jobs for the aerospace engineers related to the design and production of commercial aircraft. You also have certain areas where you can get a position in the aerospace field usually around manufacturing and cities.

The Federal Government traditionally has been the aerospace industry's biggest customer. NASA also is a major purchaser of the industry's products and services, mainly for space vehicles and launch services. The aerospace industry is dominated by a few large firms that contract to produce aircraft with Government and private businesses, usually airline and cargo transportation companies. Government purchases are largely related to defense. 63 percent of the jobs in aerospace manufacturing were in large establishments that employed 1,000 or more workers.

Aerospace manufacturing provided 444,000 wage and salary jobs in 2004. The largest numbers of aerospace jobs were in Washington and

California, although many also were located in Kansas, Texas, Connecticut, and Arizona."

Biomedical engineering is one of the fastest growing fields of engineering. Employment of biomedical engineers is expected to grow faster than the average for all engineering occupations through 2012. The aging population and the focus on health issues will increase the demand for better medical devices and equipment designed by biomedical engineers.

Biological scientists, a form of biomedical engineering, held about 77,000 jobs in 2004. Slightly more than half of all biological scientists were employed by Federal, State, and local governments. Federal biological scientists worked mainly for the U.S. Departments of Agriculture, Interior, and Defense and for the National Institutes of Health. Most of the rest worked in scientific research and testing laboratories, the pharmaceutical and medicine manufacturing industry, or hospitals".

Computer hardware and software engineers may face competition for jobs because the number of degrees granted in this field has increased rapidly. Employment is expected to grow more slowly than average, and there will be intense foreign competition from the Asian countries.

"Computer software engineers held about 800,000 jobs in 2004. Approximately 460,000 were computer applications software engineers, and around 340,000 were computer systems software engineers. Although they are employed in most industries, the largest concentration of computer software engineers—almost 30 percent—are in computer systems design and related services. Many computer software engineers also work for establishments in other industries, such as software publishers, government agencies, manufacturers of computers and related electronic equipment, and management of companies and enterprises.

An increasing number of computer software engineers are employed on a temporary or contract basis, with many being self-employed, working independently as consultants. Some consultants work for firms that specialize in developing and maintaining client companies' Web sites and intranets. About 23,000 computer software engineers were self-employed in 2004.

Environmental engineers should have above average job opportunities; employment is expected to increase much faster than the average for all engineering occupations. Much of the expected growth will be due to it being recognized as a specialty rather than as an area that other engineers such as civil engineers specialize in. More environmental engineers will be needed to comply with environmental regulations and to develop methods of cleaning up existing hazards.

Urban and regional planners, a form of environmental engineer, held about 32,000 jobs in 2004. About 7 out of 10 were employed by local governments. Companies involved with architectural, engineering, and related services, as well as management, scientific, and technical consulting services, employ an increasing proportion of planners in the private sector. Others are employed in State government agencies dealing with housing, transportation, or environmental protection and a small number work for the Federal Government.

Overall employment for industrial engineers is projected to grow about as fast as the average. In addition, many openings will be created by the need to replace industrial engineers who transfer to other occupations or leave the labor force.

Operations research analysts, a sector of industrial engineering, held about 58,000 jobs in 2004. Major employers include computer systems design firms; insurance carriers and other financial institutions; telecommunications companies; management, scientific, and technical consulting services firms; and Federal, State, and local governments. More than 4 out of 5 operations research analysts in the Federal Government work for the Department of Defense, and many in private industry work directly or indirectly on national defense.

Employment of materials science engineers is expected to grow more slowly than the average although more materials science engineers will be needed to develop new materials for electronics, biotechnology, and plastics products.

Chemists and materials scientists held about 90,000 jobs in 2004. About 43 percent of all chemists and material scientists are employed in manufacturing firms—mostly in the chemical manufacturing industry, which includes firms that produce plastics and synthetic materials, drugs, soaps and cleaners, pesticides and fertilizers, paint, industrial organic chemicals, and other chemical products. About 15 percent of chemists and material scientists work in scientific research and development services; 12 percent work in architectural, engineering, and related services. In addition, thousands of people with a background in chemistry and materials science hold teaching positions in high schools and in colleges and universities. Chemists and materials scientists are employed in all parts of the country, but they are mainly concentrated in large industrial areas.

Average employment opportunities are expected in mining and geological engineering. A significant number of mining engineers currently employed are approaching retirement age, which should create new job opportunities. There are big opportunities for those looking to live abroad.

Good opportunities should exist for nuclear engineers because the small number of nuclear engineering graduates will likely to be in balance with the number of job openings. Nuclear engendering has a lot of growth potential. A lot of funds have gone into nuclear power. Employment of petroleum engineers is expected to decline through 2012 because most of the potential petroleum producing areas in the United States has already been explored. With the discovery of large oil deposits located in the Gulf Coast, the need for Petroleum Engineers may increase over the coming years."

Four out of ten engineering positions are found in manufacturing industries such as transportation and equipment manufacturing and computer and electronic product manufacturing.

The next table shows current and projected employment for engineers, according to a 2007 survey by the National Association of Colleges and Employers.

Projections data from the National Employment Matrix

Occupational title	Employment, 2006	Projected employment, 2016	Change, 2006-16	
			Number	Percent
Engineers	1,512,000	1,671,000	160,000	11
Aerospace engineers	90,000	99,000	9,200	10
Agricultural engineers	3,100	3,400	300	9
Biomedical engineers	14,000	17,000	3,000	21
Chemical engineers	30,000	33,000	2,400	8
Civil engineers	256,000	302,000	46,000	18
Computer hardware engineers	79,000	82,000	3,600	5
Electrical and electronics engineers	291,000	306,000	15,000	4
Electrical engineers	153,000	163,000	9,600	6
Electronics engineers, except computer	138,000	143,000	5,100	4
Environmental engineers	54,000	68,000	14,000	25
Industrial engineers, including health and safety	227,000	270,000	43,000	19
Health and safety engineers, except mining safety engineers and inspectors	25,000	28,000	2,400	10
Industrial engineers	201,000	242,000	41,000	20
Marine engineers and naval architects	9,200	10,000	1,000	11
Materials engineers	22,000	22,000	900	4
Mechanical engineers	226,000	235,000	9,400	4
Mining & geological engineers, including mining safety engineers	7,100	7,800	700	10
Nuclear engineers	15,000	16,000	1,100	7
Petroleum engineers	17,000	18,000	900	5
Engineers, all other	170,000	180,000	9,400	6

Earnings for engineers vary significantly by specialty, industry, and education. Variation in median earnings and in the earnings distributions for engineers in various specialties is especially significant. The table below shows average starting salary offers for engineers, according to a 2007 survey by the National Association of Colleges and Employers.

Earnings distribution by engineering specialty, May 2006				
Specialty	Lowest 10%	Lowest 25%	Median	Highest 25%	Highest 10%
Aerospace engineers	59,610	71,360	87,610	106,450	124,550
Agricultural engineers	42,390	53,040	66,030	80,370	96,270
Biomedical engineers	44,930	56,420	73,930	93,420	116,330
Chemical engineers	50,060	62,410	78,860	98,100	118,670
Civil engineers	44,810	54,520	68,600	86,260	104,420
Computer hardware engineers	53,910	69,500	88,450	111,030	135,260
Electrical engineers	49,120	60,640	81,050	99,630	119,900
Environmental engineers	43,180	54,150	69,940	88,480	106,230
Health and safety engineers, except mining safety engineers and inspectors	41,050	51,630	66,290	83,240	100,160
Industrial engineers	44,790	55,060	68,620	84,850	100,980
Marine engineers and naval architects	45,200	56,280	72,990	90,210	113,320
Materials engineers	46,120	57,850	73,990	92,210	112,140
Mechanical engineers	45,170	55,420	69,850	87,550	104,900
Mining and geological engineers, including mining safety engineers	65,220	77,920	90,220	105,710	124,510
Nuclear engineers	65,220	77,920	90,220	105,710	124,510
Petroleum engineers	57,960	75,880	98,380	123,130	over 145,600
All other engineers	46,080	62,710	81,660	100,320	120,610

In the Federal Government, mean annual salaries for engineers ranged from \$75,144 in agricultural engineering to \$107,546 in ceramic engineering in 2007.

As a group, engineers earn some of the highest average starting salaries among those holding bachelor's degrees.

About 555,000 engineering jobs were found in manufacturing industries, and another 378,000 wage and salary jobs were in the professional, scientific, and technical services sector, primarily in architectural, engineering, and related services and in scientific research and development services. Many engineers also worked in the construction and transportation, telecommunications, and utilities industries.

Federal, State, and local governments employed about 194,000 engineers in 2004. About 91,000 of these were in the Federal Government, mainly in the U.S. Departments of Defense, Transportation, Agriculture, Interior, and Energy and in the National Aeronautics and Space Administration. Most engineers in State and local government agencies worked in highway and public works departments. In 2004, about 41,000 engineers were self-employed, many as consultants.

Engineering Positions

Typically, new graduates begin as trainees, junior designers or manufacturing and test engineers. Typically, you are given the title of Junior Engineer. In smaller companies where the projects don't require a lot of time and manpower, you may be promoted to a Project Engineer and ultimately a Project Manager. In larger companies the progression is more traditional, moving from Junior Engineer to Intermediate Engineer, to Senior Engineer before becoming a Project Manager. Project Managers are usually less technical and are more involved with the overall management of the project from the point of view of maintaining goals and schedules.

You may then be assigned to sales and marketing or get into general engineering management. A large percentage of engineers no longer do engineering work by the tenth year of their employment. At that point they often advance to supervisory or management positions.

An MBA enhances your opportunities for promotion. A doctoral degree is essential for university teaching or supervisory research positions. When you get to this point you may decide to start your own consulting firm. Continued advancement, raises, and increased responsibility are not automatic but depend on sustained demonstration of leadership abilities.

Work Place

Most engineers work in office buildings, laboratories, or industrial plants. Others may spend time outdoors at construction sites or oil and gas exploration and production sites, where they monitor or direct operations or solve onsite problems. Some engineers travel extensively to plants or worksites. Many engineers work standard 40 hour weeks but most are project oriented and have a fixed time frame for their team to complete their project and are required to put in whatever number of hours it takes to that end.

Federal, state, and local governments employ many engineers. Most are in the Federal Government, mainly in the U.S. department of Defense, Transportation, Agriculture, Interior and Energy. Those employed at the state and local government levels, work in highway and public works departments.

About half of all engineers work for companies that design and manufacture electronics, scientific instruments, automobiles, aircraft, chemicals or petroleum products. Biomedical engineers work at universities that are affiliated with hospitals or medical complexes. The remaining group is divided between public utilities and federal, state and local government departments and agencies, from NASA to the state highway department.

More than 40 percent of all civil engineers work for the government. Another 3 percent are in private engineering or architectural firms. Opportunities tend to vary by geographic area and with the health of the economy.

About two thirds of all chemical engineers are employed in manufacturing industries such as chemical, petroleum refining and paper or plastics. Most others work for engineering services, research, testing services or consulting firms that design chemical plants. Some work for government agencies or as independent consultants.

Salaries

Earnings for engineers vary significantly by specialty, industry, and education. Even so, as a group, engineers earn some of the highest average starting salaries among those holding bachelor's degrees. The following tabulation shows average starting salary offers for engineers, according to a 2005 survey by the National Association of Colleges and Employers. This inflated entrance salary places pressure on internal salary levels for companies that recruit many new graduates. Often an experienced engineer makes only a few thousand more than a recent graduate and salary growth tends to remain sluggish after that.

Average starting salary by engineering specialty and degree , 2007

Curriculum	Bachelor's	Master's	Ph.D.
Aerospace/aeronautical/ astronautical	$53,408	$62,459	$73,814
Agricultural	$49,408	—	—
Architectural	$48,664	—	—
Bioengineering and biomedical	$51,356	$52,240	—
Chemical	$59,361	$68,561	$73,667
Civil	$48,509	$48,280	$62,275
Computer	$56,201	$60,000	$92,500
Electrical/electronics and communications	$55,292	$66,309	$75,982
Environmental/ environmental health	$47,960	—	—
Industrial/manufacturing	$55,067	$64,759	$77,364
Materials	$56,233	—	—
Mechanical	$54,128	$62,798	$72,763
Mining & mineral	$54,381	—	—
Nuclear	$56,587	$59,167	—
Petroleum	$60,718	$57,000	—

Footnotes:

(NOTE) Source: National Association of Colleges and Employers

Five to eight years into your career, your compensation as a project manager is usually in the $85,000 range. Lab engineers and senior managers top around $120,000. Even at the best paying companies, engineers with no management responsibilities don't earn more than $150,000. Some companies offer a dual career path. Exceptional engineers will have salaries comparable to senior managers but will focus on research, design and invention without the administration responsibly of managers. Partners in engineering, architectural or consulting firms earn from $100,000 to $150,000 or more. If you pursue a management track you can earn considerably more. A corporate vice president of engineering can earn a salary of more that $200,000 at a technical company. Many companies offer stock options as an incentive for keeping good engineers, while holding down their salary costs. This is a double edged sword. While some engineers become millionaires through their stock options, at other companies, the stock may become worthless after a period of time.

In addition to salaries, most private companies and many public companies offer stock options, medical, retirement plans and have yearly bonus plans. Stock options alone create more millionaires among engineers than anything else.

Informational on how to find an Engineering Position

Job Sites
• www.AIChE.org – This is the American Institute of Chemical Engineering (AIChE), an excellent site. The site contains lots of professional information (conferences, publications, education & training, etc.) in addition to the Careers & Education section which contains some job postings as well as an extensive list of links to job opportunities for chemical engineers.

• www.engineerjobs.com – At the Engineer Job Source you can edit and delete your resume, if you post one, and you can also disguise your identity relatively easily. Engineering job listings are available by state or by job title. Jobs are available through RSS feed in several engineering job categories.

• www.environmentalcareer.info/index.asp – There you can browse international environmental job listings and also browse through the event calendar.

• www.ecoemploy.com – The Environmental Jobs and Careers organization has environmental job postings (for the U.S. and Canada), links to government environmental job openings, recruiters, and employers by state or province. There are also links to international resources, government resources in the U.S. and Canada, etc.

• careers.ieee.org – From the Institute of Electrical and Electronics Engineers, job listings and career resources for engineers are posted.

• www.waterjobnetwork.net – Water Jobs Now is a job site for the drinking water community: treatment plant operators and managers, engineers, scientists, environmentalists, etc. interested in water supply and public health. Search for a job by location in the U.S., category (e.g. conservation, engineer, finance, etc.), job setting (consultant, government, utility, etc.), and date posted.

• Other, more general technology employment websites include www.careerbuilder.com and www.monster.com.

Professional and Industry Associations and Societies may also contain job openings, lists of potential employers, chapter meetings where you can connect with a potential coworker or employer, and news about your profession or industry so you can stay "current" even if you are not employed. Here is a list of associations:

• www.aaai.org – American Association for Artificial Intelligence
• www.aaes.org – American Association of Engineering Societies
• www.astronautical.org – American Astronautical Society
• www.acs.org – American Chemical Society
• www.acec.org – American Council of Engineering Companies
• www.adda.org – American Design Drafting Association
• earth.agu.org – American Geophysical Union
• www.aiaa.org – American Institute of Aeronautics and Astronautics
• www.aia.org – American Institute of Architects
• www.aiche.org – American Institute of Chemical Engineers

- www.aip.org – American Institute of Physics
- www.ams.org – American Mathematical Society
- www.aps.org – American Physical Society
- www.asee.or – American Society for Engineering Education
- www.asabe.org – American Society of Agricultural and Biological Engineers
- www.asce.org – American Society of Civil Engineers
- www.ashrae.org – American Society of Heating, Refrigerating and Air-Conditioning Engineers
- www.asme.org – American Society of Mechanical Engineers
- www.asm-intl.org – American Society of Microbiology
- www.navalengineers.org – American Society of Naval Engineers
- www.asse.org – American Society of Safety Engineers
- www.apwa.net – American Public Works Association
- www.acse.org – Association of Chinese Scientists and Engineers
- www.acm.org – Association for Computing Machinery
- www.crsi.org – Concrete Reinforcing Steel Institute
- www.geosociety.org – Geological Society of America
- www.iesna.org – Illuminating Engineering Society of North America
- www.ieee.org – Institute of Electrical and Electronics Engineers
- www.iienet2.org – Institute of Industrial Engineers
- www.ite.org – Institute of Transportation Engineers
- www.tms.org – Minerals, Metals, and Materials Society
- national.nsbe.org – National Society of Black Engineers
- www.nspe.org – National Society of Professional Engineers
- www.sampe.org – Society for the Advancement of Material and Process Engineering
- www.same.org – Society of American Military Engineers
- www.sae.org – Society of Automotive Engineers
- www.scs.org – Society for Computer Simulation
- www.siam.org – Society for Industrial and Applied Mathematics
- www.sme.org – Society of Manufacturing Engineers
- www.swe.org – Society of Women Engineers

Engineering Jobs in the Federal Government. Many Federal Agencies Offer Attractive Bonuses and Accelerated Promotions. Federal agencies are increasingly offering bonuses and promotion programs for new employees that rival those of the private sector.

• Recruitment Bonuses: Agencies may pay a recruitment bonus of up to 25% of the base salary to a new employee if the position is hard to fill.

• Relocation Bonuses: Agencies may pay a relocation bonus of up to 25% of base pay to an employee who must relocate for the position

• Accelerated Promotions: Agencies may offer a one-time accelerated promotion for entry level hires

Examples of these types of benefits in practice include:
• The National Nuclear Security Administration's Future Leaders Program offers a recruitment bonus of $6,000 and GS employees may receive an accelerated promotion after the first six months of employment.

• At the U.S. Patent and Trademark Office new entry level Patent Examiners are eligible for a one-time accelerated promotion after six months.

• Candidates who enter the Professional Development Program with the Federal Highway Administration will be offered travel and relocation assistance to cover their initial move. Recipients of this assistance must agree to commit a minimum of one year of service.

Job Profiles
Explore more and apply now at www.usajobs.com. Here are examples of recent job openings.

Aerospace Technology (AST) – General Engineer at NASA
• Salary Range: $42,290.00 – $56,246.00
• Duty Locations: Houston, TX; Las Cruces, NM; Kennedy Space Center, FL; Russia
• Summary: The National Aeronautics and Space Administration (NASA), the world's leader in space and aeronautics is always seeking outstanding scientists, engineers, and other talented professionals to carry forward the great discovery process that its mission demands.
• Major Duties: Coordinate and collaborate with others such as project managers, research engineers, hardware engineers, managers, and others on standard design or simulation requirements. Participate in the

specifications, acceptance testing, and evaluation of technology, equipment, or systems components. Serve as project engineer for projects of limited scope, such as those that require minor adaptation of existing methods and techniques.

Electronics Engineer at the Department of Transportation

- Salary Range: $38,700.00 – $89,300.00
- Duty Locations: Many vacancies throughout the nation
- Summary: Join the U.S. Department of Transportation in meeting the challenges of the 21st Century, shaping the future of the organization, and advancing the best transportation system in the World.
- Major Duties: Install and maintain electronic equipment and lighting aids associated with facilities and services required for aviation navigation to assure a reliable, safe, and smooth flow of air traffic. Assist in the design, development and evaluation of new types of electronic equipment.

Civil Engineer at the Department of Energy

- Salary Range: $68,625.00 – $89,217.00
- Duty Locations: New York City, NY and Chicago, IL
- Summary: Winning more R&D awards than any private sector organization and twice as many as all other federal agencies combined, the U.S. Department of Energy is the nation's top sponsor of research and development in fields such as alternate fuel vehicles, energy efficiency, gene research, supercomputers and microelectronics.
- Major Duties: Inspect engineering evaluation and analysis of non-federal hydroelectric projects to insure the safety of the projects and compliance with the terms of the license, the provisions of the Federal Power Act and the Commission regulations. Review and perform evaluations and analyses; and prepare engineering reports to establish and document the dam safety, public safety, and environmental and license compliance of the Commission hydroelectric projects.

Patent Examiner (Electrical and Computer Engineering) at the Patent and Trademark Office

- Salary Range: $40,200 – $74,500
- Duty Locations: Alexandria, VA
- Summary: Protect intellectual property rights and prevent products from copyright infringement or forgery. Patent and Trademark Office Patent Examiners review patent applications and assess patentability of inventions.
- Major Duties: Review patent applications to determine whether they comply with basic format and legal requirements, research similar prior inventions and technologies and compare to invention claimed in new patent application, communicate examination findings to inventors.

Mechanical Engineer at the Department of the Army

- Salary Range: $51,271 – $65,704
- Duty Locations: Ft Riley & Manhattan, KS
- Summary: Civilian employees serve a vital role in supporting the Department of the Army's mission. They provide the skills that are not readily available in the military, but crucial to support military operations. Army Corps of Engineers of the Army integrates the talents and skills of its military and civilian members to form a Total Army.
- Major Duties: Prepare various types of cost estimates to serve as a basis for positions relative to change orders and/or modification actions. Analyze and break down plans and shop drawings. Compare cost estimates with contractor's proposals and analyzes differences. Negotiate agreements for contractual effect of change orders or modifications. Make engineering studies and analysis of recommendations.

Job Opportunities at National Laboratories

In addition to many job opportunities in the federal government agencies, students with advanced degrees in engineering can also find opportunities at National Laboratories. National Laboratories are funded by the government (e.g., Department of Energy) but operated by non-governmental private firms or universities. They are world-renowned research and development centers with a tradition of pursuing knowledge at the frontiers of science and a long history of excellence in a range of fields, including the basic sciences, applied energy research, and weapons-related technologies. Below are the major National Laboratories. Visit the respective Web sites for information on job and career opportunities.

Major National Laboratories

- www.external.ameslab.gov – Ames Laboratory
- www.anl.gov – Argonne National Laboratory
- www.bnl.gov – Brookhaven National Laboratory
- www.fnal.gov – Fermi National Accelerator Laboratory
- www.inel.gov – Idaho National Engineering and Environmental Laboratory
- www.jpl.nasa.gov – Jet Propulsion Laboratory
- www.lbl.gov – Lawrence Berkeley National Laboratory
- www.llnl.gov – Lawrence Livermore National Laboratory
- www.lanl.gov – Los Alamos National Laboratory
- www.nrel.gov – National Renewable Energy Laboratory
- www.netl.doe.gov – National Energy Technology Laboratory
- www.ornl.gov – Oak Ridge National Laboratory
- www.pnl.gov – Pacific Northwest National Laboratory
- www.pppl.gov – Princeton Plasma Physics Laboratory
- www.sandia.gov – Sandia National Laboratories
- www.slac.stanford.edu – Stanford Linear Accelerator Center
- www.jlab.org – Thomas Jefferson National Accelerator Facility

How to Find a Mechanical Engineering Job

Engineering is one of the hottest majors on college campuses today. The allure of high starting salaries and being in demand among employers is great. Mechanical engineers are sought after in today's job market. This demand is fueled by the continuing development of and reliance on increasingly complicated forms of technology in society. For those with a degree in this area, it isn't hard to find mechanical engineering jobs.

Finding Mechanical Engineering Jobs

Step 1

Attend job fairs. If you are in college, the engineering department at your school will almost certainly hold one or more job fairs each year. These fairs are excellent opportunities to find out what companies are hiring, what your job would involve and any special qualifications you might need. You can also talk to the people who will be doing the hiring, which gives you the chance to impress them in person.

Contact your school professors and ask if they have any employment leads or advice. Many graduates keep in touch with their professors and ask for assistance when they have a job opening.

Step 2

Obtain an internship. Internships in mechanical engineering provide invaluable experience that you can use when you start your career. Internships also allow you to make personal contacts in the industry and to establish relationships at the firm where you are interning. You can use these contacts and relationships to secure a job for yourself after graduation. In fact, many companies hire their interns outright after those interns graduate.

Step 3

Network with engineering professionals. Join local engineering groups and attend their meetings. This gives you the opportunity to meet others who are working in the field. These personal contacts can lead to job offers down the road.

Step 4

Look in the classified ads in the newspapers of the cities you want to work. Engineering firms do place want ads, just like other companies. The job you want may be listed today. If they are not, keep looking until you see something that appeals to you. New engineering jobs open up every day in this country.

Step 5

Make use of all the employment websites and professional organization websites listed on the previous pages.

Tips & Warnings

- Weigh several job offers before accepting one. You'll want to compare employers carefully. Whether an employer will pay for graduate school and whether they will require you to relocate are some important considerations. Depending on your circumstances, you may also want to go with an employer who will not require extensive travel. Of course, starting salary is also important to factor in when selecting a mechanical engineering job.

- Find a headhunter to help. Headhunters will learn what you are looking for in a job and may be able to help you. They do not get paid unless you get a job, and when they are paid, they are paid by the company, not you. Remember though that headhunters don't fine jobs for people, but rather find people for jobs and your resume may sit at the agency for a long time before it's matched up with a particular job requirement. On the other hand, temporary agencies do find jobs for people. They will enable you to network with other engineers and you will have something to put on your resume. You will get a feel for the working environment and learn about the engineering business. Many temporary employees are converted to regular employees so becoming a temp allows you to test drive a particular company that you may be interested in.

- Go to as many business related social events to network. Tell everyone that you are currently looking for a job.

- Consider being an unpaid intern. This will help you get your foot in the door, give you something to put on your resume, network with other engineers, and enable you to learn about that business. Interns are often given a full time job after their internship is over.

- While looking for a job, consider getting an advanced degree. You can get a masters degree so you can teach at a college level. Many schools now offer online classes.

- Consider job hunting a full time job. Do not expect a new job to fall in your lap. You have to work for it.

How to do a Job Search

If you're currently employed and plan to change jobs soon, don't up and quit without a plan. You don't want to find yourself jobless with no prospects if you can help it — regardless of the economy. Employed or not, be aggressive in your tactics.

The market is tougher. There are fewer jobs, more candidates and hiring authorities are being more selective. So you have to really distinguish yourself even more from the competition. You must go above and beyond the average interview.

Job hunting is about getting noticed by employers. You don't want to blend in with every other person who responds to a job posting or walks into an interview. That's as true now as it was a decade ago. If you're a good employee, you'll be a good addition to the team -- but they'll never know if you're just another faceless name in a pile of resumes.

To stay ahead of the pack use a search engine and social networking sites. If you have mutual contacts, drop their names during conversation so you become memorable. If you're lucky enough to get an interview, be just as prepared.

Carry a portfolio of reports you have written demonstrating your skills or a 30-60-90 day plan as to what you would do the first 90 days of your employment. Do extra research on the company and the person you are interviewing, and maybe speak to their customers and find out how they are perceived.

Because companies don't have the budgets they had a year or two ago, they can't afford to waste time or money on finding a replacement that's anything less than perfect — or at least close to perfect. Many employers aren't replacing vacant positions that aren't vital to their operations. If they're willing to spend on a new hire, they want a qualified candidate who will stick around for awhile. They also know that they have many job seekers for far fewer positions. The pressure is on you to be the best potential employee they'll come across in the hiring process.

Everything, and I mean everything, in your interview matters — your dress, your speech, your manners — and employers can be very unforgiving in this market, especially when they still have plenty of candidates to choose from. Remember to sit straight leaning a little forward. Do not slouch. Wait for questions to be completed. Do not interrupt the interviewer.

A resume with typos or unprofessional attire in an interview rarely bodes well for a job seeker. In 2009 such a misstep is guaranteed to get your name crossed off the list of potential candidates. Here are some things to consider during your 2009 job hunt.

Concerning your Resume avoid typos. This seems like redundant advice, but hiring managers repeatedly cite typographical errors as a top pet peeve. You can't control whether a hiring manager ever picks up your resume, whether your personality clicks with his or hers or whether you ultimately will get the job. Conversely, your resume is your creation and your first step into the company's door. You went out of your way to type it up and send it to the company. What kind of message are you sending if you don't take responsibility for one of the few factors entirely within you control? The goal of your resume is to get that first interview. Prepare more than one resume. Tailor each to the position you are applying for. Don't spend a lot of time writing about your duties and responsibilities. For a particular position, the duties and responsibilities are usually similar. Think instead in terms of your accomplishments in the positions you held. Use sentences like, "I designed a doohickey which saved my company $100,000 during the first year with a projected savings of $500,000 over the doohicky's life."

The interview is a two-way street, where you need to sell yourself to the hiring manager and he or she needs to sell the company to you. Let the company do its part and focus on yours. You always want to prove to the employer that you're looking for longevity -- in a competitive job market, it's vital. Explain that a position where you can learn, grow and be a team member for longer than a few months is your ideal situation. If the hiring manager gets the feeling that you're desperate to find any job just to earn a paycheck, you'll be out the door before you set your bag down. Employers don't want to spend the money training someone they'll be replacing in four months.

Don't get lazy. Browse job boards, search the classifieds, walk around the neighborhood, look for jobs wherever you can. Some employers don't want to spend a lot of money advertising a job opening, so reach out to companies that might not have a job opening listed, as they might be quietly searching for new employees.

Network your connections, both social and professional. Networking is an invaluable resources during a job hunt. Even friends of friends you've only met at a cocktail party are worth touching base with during a job hunt. When you let people know that you're looking for a new job, they'll keep you in mind if they run across an open position at their workplaces or if they hear about one at a friend's company. You can cover more ground than if you search alone.

Conclusion

This chapter explained how to get the position that is right for you. You now have a good idea of what type of engineering suits you and what the requirements are. It pointed out that the first step in finding a position is networking. Rules and etiquette involved in networking were defined. Also recruiters and how to present yourself was discussed. Alternative position finders such as summer employment, work study programs and cooperative experiences and others were outlined.

The general expected growth for each type of engineering was outlined giving you a detailed expectation of what you will encounter when going into any field of engineering. Traditional engineering programs such as chemical, civil, electrical and mechanical were reviewed. Specialties like aerospace, biomedical, computer hardware and software, environmental, industrial, materials science, mining, geological and nuclear engineering were outlined and now you know the expected growth and employment opportunities for each. Outsourcing was also discussed.

A new graduate's entry level position was described and now you have a good idea of what to expect in your first position. Working conditions, locations and opportunities were given in reference to the engineering fields.

The latest salaries for all types of engineers were stated and you now have a good idea of what you would be earning at different points throughout your career. It is interesting to note that often an experienced engineer makes only a few thousand dollars more than a recent graduate. You now know what areas of the engineering profession make the most and the least.

An extensive list of job related websites, professional engineering organizations, and major government funded laboratories were listed and lastly, your actual job search was covered in depth.

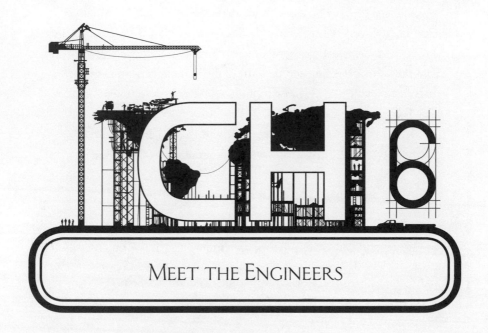

Meet the Engineers

I Am the Very Model of an Engineering Graduate
by Gary Friedman

I am the very model of an engineering graduate;
I work with systems and with problems both of nature delicate;
I've studied properties of things in motion for the hell of it,
Regurgitating answers to insure that my degree I'll get;
I've studied chemistry and learned the formulas of saturate
Solutions in normality in labs that are immaculate,
And programmed huge machines with large routines I've written just so that
Infrequently will I encounter problems that I can't attack;
I'm very good at systems both in digital and analogue;
Have studied great philosophers; can quote to you their dialogue;
I'll work all day and never quit, that is if I can manage it;
I am the very model of an engineering graduate.

I use my Hewlett-Packard for the answers found in calculus,
And problems most encountered in numerical analysis;
It calculates proportions used in heart and lung dialysis;
Eventually I'll work for them and move to where Corvallis is.
My interests are much greater than my friends and colleagues might believe;
I've worked with magic and performed illusions written to deceive;
In elementary schools I'll work with children who do not receive
The help and dedication they require so they might achieve.
My years of work and study have allowed me to become involved
In engineering problems that have only partially been solved.
My grades, however, are the pits, and I don't really give a s---,
That's why I am the model of an engineering graduate.

This is the chapter that will tell you about engineers and their lives. There are many stereotypes of engineers. Engineers are depicted as the nerds with the pens and calculators in their pockets, the solitary loners or the wiz kids. There are many different types of engineers just as there are many different types of people. Certain personality traits and skills lend themselves better to certain types of engineers.

The engineers being discussed in this chapter work in very different industries, positions and places. How they got there will intrigue you. You can find yourself in any of these situations and for many different reasons. Their stories may help you find your best place in the engineering field.

David's typical day of engineering...

David considers it a blessing that quite to the contrary, he has no typical day plan, workload, routine, or coworkers. As a young engineer working for a musical instrument manufacture in the North East, his job duties are neither well defined nor are they consistent. In the nine months he has been employed he has done everything from making design changes to the instruments themselves to weighing instruments for the purposes of updating company literature.

Working out of an office on the factory floor he is somewhat isolated from the other engineers they employ and his boss, the head of engineering. As a result he works very autonomously, rarely collaborating or working alongside other engineers. Most of his workload comes via his boss with whom he tries to confer with at least once a day, but he sometimes take on jobs at the request of a foreman, a department head, or to address something that concerns him.

He is constantly walking around the factory and the office tracking test instruments and conferring with foreman, operators, woodworkers, maintenance men, machinist, and other engineers. Most of the other engineers have less flexible schedules than he does. Two mechanical engineers that he works with often spend the bulk of their days in front of a computer modeling piano parts using Unigraphics design software. Others assume somewhat managerial roles, spending at least significant portion of their day managing key departments in the factory. David works 8:30 to 4:30 Monday threw Friday.

The Opinionated Engineer

Leah is a senior manufacturing engineer. She supports existing chemical processes and equipment in making automotive brake components in the plant. She troubleshoots problems, works on maintenance to guide them with repairs. She works with the analytical people to determine the correct chemical mixtures for the company's many products, she also works with the quality control people to insure that parts are manufactured to specification, and looks for ways to improve existing processes and reduce costs.

Leah has been working in the business for 13 years. She has a BS in Chemical Engineering and an MS in Materials Science Engineering. She does not have a professional certificate because it is not required by her company, or in her engineering field in Ohio. She typically works a 45 to 50 hour week, occasionally working weekends when needed. She comes in at 6:30 in the morning and leaves at 4 pm.

On a typical day she begins by listening to her voice mail, reads e-mail and makes her rounds of the various manufacturing and fabrication areas within her building. Who or what she runs into usually determines how long it takes her to complete her rounds. For example, maintenance folks will stop her to show her a problem or ask her to order parts. Supervisors may stop her to explain quality issues or equipment problems they are having.

When she gets back to her desk, she replies to her phone messages, writes specifications or purchase orders, calls suppliers inquiring about parts that are due to be delivered or how something works and talks with other engineers about electrical problems or changes needed. She may also work with lab analysts on chemical issues and write memos to her manager about issues that have come up recently. She is always ready to jump up to go look at something the minute there is a problem on the production floor.

In her 13 years with the company she has advanced two times up the manufacturing engineering ladder. These advances have resulted in more responsibilities due to her increased level of experience and competence. Further movement would move her into managing other engineers and doing less actual hands on engineering work. She can't see herself moving up for some time because she values her free time too much. She doesn't wish to spend any more time at work. She told this to her superiors and they seemed to be okay with it. She usually has lunch at her desk while working, sometimes while running errands. Her afternoon is usually a continuation of her morning.

She gets along well with her fellow workers and inquires about her co-workers' outside interests and families. She is open minded and willing to listen to other people's ideas and suggestions. She feels there are many ways to solve a problem and people will often approach things from different perspectives. She doesn't tell everyone that her way is the only way every time. She is technically competent and resourceful and she usually knows the answer to problems but she also asks experienced people for input. She is credible and she feels this is critical in her position.

She likes the plant environment and the daily variety in her job. She never finds it monotonous. She has a lot of short term goals that she can accomplish. Every time a piece of equipment stops working she really feels great when she can get it back up and running again.

She dislikes the bureaucracy of the manufacturing union that the actual production workers are in nor does she like the bureaucracy associated with upper management at her company. It is a large company and the many levels of management can be exasperating and cumbersome. Also the strong manufacturing workers union at the facility makes it seem like management and the union are often butting heads and do not have the same goals. She gets frustrated with people who do not work as hard as she does. She feels there is a lot of fat in certain areas of the organization.

She says the biggest misconceptions about the engineering field are that engineers make more money than the average and also that you should be able to do any type of engineering work if you have an engineering degree. Having an engineering degree won't guarantee that you'll be good at every aspect of engineering. In fact, she feels you use very little of what you learn in college on a day to day basis.

She works for a company with 2,000 plus employees with locations in most countries. Her salary is about $80,000.

The Partner Personality

Jack is feeling more at ease because a major contract his new firm undertook is almost complete. He is one of the original partners in the firm and he has had many anxious times in the past few years waited for the firm's business to ramp up. Now with the completion of the biggest project almost done he can breathe a little easier.

Jack's firm specializes in site preparations. It does surveys to determine the size and shape of the terrain at a project site, as well as underground probes to find where bedrock and underground streams might be.

A new hospital is being built in a rather remote corner of a valley. The general contractor gave Jack's firm a contract to do the site preparation. Jack was the overall supervisor for his firm's work.

The first thing he did was a site survey, which consisted of carrying electronic surveying equipment up and down the hills on the site. Jack then hired a photogrammetry expert to take aerial photographs and to make drawings of the site elevations. He spent a good part of the spring hiking with the surveying crew monitoring their performance.

After that, he brought a drilling crew out to the site in order to take core samples of the ground. He monitored this work carefully and specified where the crew should dig. He also analyzed the core samples as they were delivered from the site. They did find a few underground streams and an old dumping ground. The general contractor and the hospitals management firm were extremely interested in these findings because of the new federal and state laws stating that the landowner must clean up any harmful waste products found at the site.

Jack's firm characterized the size and composition of the waste materials and the chemical laboratory tests showed what type of waste products they were. As he finished the final report of the site preparation survey he was proud of his firm and his partnership ability. His firm had handled all the details of the surveying, provided the guidelines for where earth had to be moved and saved the owners from a big problem by finding the dumpsite before construction began. Jack anticipates that this project will lead to more contracts with other general contractors.

Camp Made Engineer

Ann was exposed to engineering as a child. Her interest in engineering came from her father who was a quality control engineer. She also had the chance to learn about engineering in high school through a National Science Foundation engineering camp. Ann spent her time at the camp learning about science and research. Before that, she had the perception that engineers were male and on the slightly nerdy side. That camp changed her perception.

In the fourth grade she hated math and felt like the dumbest kid in her class. After going to engineering camp she changed her perspective of herself and what she could do. Then she started getting good grades and being interested in becoming an engineer.

During her freshman year at college she was recruited by the Society of Women Engineers and has been an active member ever since. She became president in her senior year. She says some people believe that the Society of Women Engineers is a crutch for women engineers but it is not. The reason she joined was to meet other women engineers. She has also participated in the Engineering Ambassadors program and the Honors Research program. She was chosen as an executive member of the Ambassadors her junior year and said she had a chance to plan some goals and objective for the ambassadors.

In her job as kitchen manager, she has the opportunity to incorporate total quality management into the kitchen environment. She is in charge of the facility and scheduling workers, helps train dietitians and works as a trouble shooter. "I make sure everyone does their job" she says.

One of a Kind

Bill has the opportunity to work in an exciting first of its kind artificial intelligence program to automate a common process, water treatment. His company, a contract supplier of water treatment systems wants to replace the instant attention to water conditions that requires several workers with a control computer that would inject the right mix of anti-corrosion chemicals at the right time. It's important work for an electrical engineer who just completed his master's degree.

One of the reasons bill took the job was that there were few electrical engineers at the company and no one in his specialty which is artificial intelligence. This gives him the chance to work on totally new systems. It also puts a spotlight on him because if he doesn't get the job done everyone will know it.

The system he is designing incorporates a new semiconductor chip with an off the shelf personal computer. The chip and the PC have all the intelligence needed to compute the right control actions but Bill needs to determine the right goals of the system so that the exact instructions can be programmed into the PC. He conducts expert system development which involves interviews with several senior engineers who have either a chemical or mechanical engineering background. He finds out that the chemicals change the acidity of the water, causing dissolved salts to precipitate out before the process water runs through the system. These areas have him look back at his college textbooks to review his understanding of chemistry and mechanics.

He then puts all the instructions into the computer and laboratory testing shows the system runs correctly. The system will then be field tested at a customer's site. Bill will monitor that work for the rest of the year.

Love the Sport Engineer

When Tom was in high school, he really loved math and science and he knew he wanted to go into some type of engineering. When he applied to colleges he went down the list of engineering programs and aerospace sounded the best to him even though he never was a big airplane fan. Tom worked for Martin Marietta in New Orleans as a cooperative education student following his freshman year. At the time, the company was building fuel tanks for the space shuttle. It was exciting work for him but it also was somewhat discouraging. He learned a lot but also saw that there were several hundred engineers working on the same project. He knew this type of environment was not for him.

Returning to college that Fall, Tom discovered an aspect of aerospace engineering that was more appealing to him — the lab. The first time he walked into the lab he knew he had found what he wanted to do. He wanted to do research on athletic equipment. After graduating Tom headed for graduate school but fate stepped in and he needed a summer job. His father worked for a company that sold raw materials to a big golf ball manufacturing company and suggested he send his resume there. As luck had it, the company was in the process of decentralizing its research and development operation and needed an engineer to design and test athletic equipment.

Tom was hired and he has been there ever since. Tom said the first day on the job he pulled out one of his aerodynamics textbooks and found information he needed. He feels engineering schools taught you how to solve problems but doesn't teach you everything that can provide you with a good solid foundation, so its up to you to use what you learned.

The Flying Engineer

Ralph likes to think of himself as an avionics engineer. He is part of a group of twelve electronics engineers in a section of fifty avionics engineers all of whom work for a major defense contractor. The fifty engineers are responsible for all the avionics for a new fighter jet the contractor is designing for the air force.

The project was typical for the aerospace industry because it involved competition among three defense contractors. Several months ago, after submitting designs to the air force each firm was given a sum of money and also had to invest its own money to build a prototype of the aircraft. The three prototypes were then tested competitively and the winner was chosen. Ralph's firm won.

His team now had to design the electronic subsystems that were only outlined during the competition. He is now finding that the electronic technology has changed drastically since the time when the original design proposal was written. Ralph has to decide whether to go along with the subcontractor's wish, insist on the existing design or propose something entirely different. After reading the performance criteria closely he realizes that there is a trade off. There is a new chip on the market, but it is more expensive but it can perform better. He checks with the group members responsible for this and finds that they are willing to change their design to suit his.

After several meetings they arrive at a preliminary cost and find that the costs roughly balance out. Ralph now begins a report for his group leader. His report will become part of another larger document that the group leader will send to the company's Vice President and eventually to the Defense Department for review.

Seas of Opportunities

Lee never heard of an agricultural engineer in China where he grew up. His professor at the National Taiwan University changed all that when he told Lee that if he wanted to save the world then begin as an agricultural engineer. Those words lured Lee into the program at the university where he would graduate with a bachelor's degree in agricultural engineering.

He then left to seek his fortune in the United States and continued his education at Michigan State University where he earned a master's degree and PhD in agricultural engineering. He is now a professor for biosystems engineering at the University of Hawaii.

After teaching agricultural machine design for several years Lee interests turned from land to water.

Lee received funding to start and maintain an aquaculture program at his school. Since then he has made major breakthroughs in raising shrimp and oysters for mass production and is developing some innovative uses for algae. He says he is an agricultural engineer, but what he does best is to grow things and produce things from biological systems.

Currently, Lee is working on a patent involving antibiotics made from algae. The drug is being tested on animals and could someday be used against bacteria that have become resistant to existing drugs.

Lee predicts that within the next 10 years, jobs in aquaculture will be plentiful as fish production becomes important as a food source. He believes that an agricultural and biological engineering background allows people entering this field to be versatile in these areas. He says there is never a dull moment in his career and you put the puzzle together piece by piece. He feels there will always be a need for production from biological systems and as for saving the world, he knows he is doing something useful.

The Proud Engineer

Abe is a proud engineer responsible for developing a new method of making steel called thin slab continuous casting. If Abe's project succeeds it will eliminate most of the hot strip which would drastically reduce the cost of producing sheet steel. The major problem is to produce steel with few flaws because this high quality steel will be used on the outer surfaces of products.

Abe is not alone in this project. The entire company management is watching its progress daily. Today, Abe's challenge is to improve the heat retention of the molten steel as it comes out of the furnace. If it cools too quickly, cracks form as it is being drawn into the thin slab. To research the problem, Abe set dozens of high temperature thermocouples in the channel where the molten steel flows. As it goes by he gets a set of readings of the temperature in each section of the channel. Abe finds that the middle of the channel is hot enough but that the sides are too cool. There are two choices available to him, adding heat to the sides or insulating it better so that it doesn't cool as quickly. Abe senses that insulation may be the better choice and less expensive. He now begins researching the variety of insulating bricks that are available. It will take him some time to finish his findings and decide on which to use.

Footprints in the Environment

Carl attended the University of Wyoming to study civil engineering. His father had an engineering degree and it seemed like a logical thing for Carl. When he finished his bachelor's degree he joined the Peace Corps and was sent to Africa. He had always been interested in the Peace Corps and it gave him a chance to travel and do something worthwhile. During the first year Carl worked on a small pipeline project to supply water to a community in the area. During his second year he worked with small groups of women training them to build concrete tanks to catch and store rainwater. He fell in love with working with water.

Since he was now very interested in working with water he decided to return to school to get a master's degree in environmental engineering. Upon completion of his masters he landed a position working for a consulting environmental engineering firm. He focuses mainly on projects related to the expansion of municipal wastewater treatment facilities. He learned how people and societies function through both his formal training and his activities in his college and the Peace Corps. He is known for his communication skills with people of all types.

Senior Structural Engineer

Sam is a Senior Structural Engineer his education is a BS in civil engineering, Rutgers University, and an MS civil engineering with emphasis in structures, Rutgers University

His job Description reads:

New bridge construction

Bridge renovation

Building design, construction and renovation

Structure analysis for placement of communications equipment

Current Projects: Preliminary design phase for replacement of a three-span bridge

Why I Took the Job: "Previously I had been working for a small firm that dealt only with one type of bridge design. I felt the opportunity to work at a firm with a wider range of projects would give me a better base for a structural engineering background, so I would be able to handle more phases of different kinds of projects."

How My Degree Helped Prepare Me: "By the time you've gotten out of college, you've decided which branch of civil engineering you'd like to go into. Then once you go out and get your first job, hopefully you have the chance to work with a firm that allows you to do many types of structural engineering, working with bridges and buildings... From there, you can focus even further, to bridge or building design or tower analysis for lattice towers, monopoles or transmission towers."

Non-Technical Skills Needed to Succeed: "The biggest thing is communication. That helps the engineers to be able to ask the questions they need to ask, and to be able to understand the answers that they receive. Often, when you get out of college, it's hard to explain yourself or the question you're trying to ask, and then it's difficult to understand the answer that you're getting because you don't have as much knowledge as the person you're working with. Things get busy on the job, so communications allow you to know when and how assignments should be completed."

Advice to Future Structural Engineers: "Always ask questions. Be persistent in trying to learn new ideas. The information is not always as readily available to you as it was in school, so you have to search it out. When you have the opportunity to work on a project and you're asking questions, be persistent enough to get all the information, so that the next time you can complete that assignment on your own."

What the Future Might Hold for Me: "First, I plan to get my Professional Engineering license and further my experience in the field. I think there will be opportunities for me later on to concentrate more in the bridge engineering field, to allow me eventually to take bridge jobs from start to finish."

Project Manager

Michael's education is a B.S., civil and environmental engineering, Clarkson University, M. S., civil engineering, Clarkson University

Job Description:

Building design

Oversight of three engineers and three drafters

Maintenance of project budgets and schedules

How I Knew This Was the Field for Me: "My father was a general contractor, so I used to work during the summer for him. Just by being around the building industry, I've always had a lot of satisfaction from seeing a building start from scratch and then go completely up. It's been a very gratifying experience seeing that."

How My Degree Helped Prepare Me: "A lot of the fundamental courses that you take in the civil engineering curriculum are required for structural engineering. Plus I think any engineering school, even more important than teaching you the fundamentals, teaches you how to think and how to solve problems. That's what engineering is about, solving complex puzzles and learning to use resources and call on other people if you don't know something, so you can learn how to get the job done."

Advice to Future Structural Engineers: "Make sure you know what you're getting into. I think structural engineering is a lot more technical in nature than civil engineering in general, or than most other engineering fields. Make sure you know the basics."

Biggest Job Surprise: "One of the big surprises going into the field was that even though there is oversight for our calculations, a good portion of the work goes unchecked. You need to have a fair degree of competence in what you're doing. You have to be very technically sharp."

Non-technical Skills Needed to Succeed: "You need a lot of organizational skills, and both verbal and written communications skills on a daily basis. You also need to have strong negotiation skills and leadership skills."

What the Future Might Hold: "I've just recently stepped into this project manager position, and I think upper management is where I'm heading. I think I have more opportunity going in that direction than in pursuing it from the technical end or from the field."

Conclusion

These are some of the examples of engineers. A partner engineer, camp made engineer, one of a kind engineer, love the sport engineer, and flying engineer are just a few engineering types that were described in this chapter. Their stories gave you a good idea of how and why they became engineers, the jobs they do and why they do it. You have learned what to expect in engineering situations and it gave you a realistic view of the life of different types of engineers.

The day to day schedule discussed in the opinionated engineer's story gave you a good description of what it would be like on the job as a senior manufacturing engineer. The partner personality enabled you to look at one of the original partners of a firm and what his inner feelings were. The camp made engineer describes how going to engineering camp in high school influenced her decision to become an engineer. The one of a kind engineer gave you insight on what it would be like to work on a first of its kind project. These are just a few of the highlighted engineers that were depicted in this chapter.

All of the engineers in this chapter gave you ideas to use to succeed in the engineering profession. From how to win a contract to overcoming complications to completing a project were addressed. Use what you have learned from these engineers and you will succeed in the engineering profession.

Team Work and Interdepartmental Relationships

Many relationships exist in the engineering profession. You must learn what they are and how to behave for best results. This chapter will discuss the many types of relationships and give you a better concept of how to relate to them.

Team Work

Working in a team is an important part of being an engineer. You are more likely to work as a team than by yourself. To achieve great things, you need a team. Building a winning team requires the understanding of certain principles.

Whatever your goal or project, you need to add value and invest in your team so the end product benefits from more ideas, energy, resources, and perspectives. Here are some principals that will help you have a successful team.

First, people who try to achieve great things by themselves sometimes do so because of the size of their ego, their level of insecurity, or simple naiveté and temperament. This is not the norm. In general, one is too small a number to achieve greatness. You are more successful in a team than by yourself.

The goals of the team should be more important than your role within it. Members of the team must be willing to subordinate their roles and personal agendas to support the team's vision. By seeing the big picture, effectively communicating the vision to the team, providing the needed resources, and hiring the right players, you can create a more unified team.

All team members have a place where they can add the most value. Essentially, when the right team member is in the right place, everyone benefits. To be able to put people in their proper positions and fully utilize their talents to maximize potential, you need to know your players and the team situation. If you can, evaluate each person's skills, discipline, strengths and weaknesses, emotions, and potential.

Focus on the team and the dream should take care of itself. The type of challenge determines the type of team you require: A new challenge requires a creative team. An ever-changing challenge requires a fast, flexible team. An Everest-sized challenge requires an experienced team. See who needs direction, support, coaching, or more or less responsibility. Add members and change leaders to suit the challenges of the moment, and remove ineffective members when you can.

The strength of the team is impacted by its weakest link. When a weak link remains on the team the stronger members identify the weak one and end up having to help him, come to resent him, become less effective, and ultimately question their leader's ability.

Winning teams have players who make things happen. These are the catalysts, or the get-it-done-and-then-some people who are naturally intuitive, communicative, passionate, talented, creative people who take the initiative, and are responsible, generous, and influential.

A team that embraces a vision becomes focused, energized, and confident. It knows where it's headed and why it's going there. A team should examine its moral, intuitive, historical, directional, strategic, and visionary compasses. Does the business maintain a high level of integrity? Do members stay? Does the team make positive use of anything contributed by previous teams in the organization? Does the strategy serve the vision? Is there a long-range vision to keep the team from being frustrated by short-range failures?

How is your team attitude? A 'Bad Apple' attitude ruins a team. The first place to start is with yourself. Do you think the team wouldn't be able to get along without you? Do you secretly believe that recent team successes are attributable to your personal efforts, not the work of the whole team? Do you keep score when it comes to the praise and perks handed out to other team members? Do you have a hard time admitting you made a mistake? If you answered yes to any of these questions, you need to keep your attitude in check.

Teammates must be able to count on each other when it counts. Is your integrity unquestionable? Do you perform your work with excellence? Are you dedicated to the team's success? Can people depend on you? Do your actions bring the team together or rip it apart?

The team fails to reach its potential when it fails to pay the price. Sacrifice, time commitment, personal development, and unselfishness are part of the price we pay for team success. The team can make adjustments when it knows where it stands. It is essential to evaluate performance at any given time, and it is vital to decision-making. Any team that wants to excel must have good substitutes as well as starters. The key is to continually improve the team.

The type of values you choose for the team will attract the type of members you need. Values give the team a unique identity to its members, potential recruits, clients, and the public. Values must be constantly stated and restated, practiced, and institutionalized.

Effective teams have teammates who are constantly talking and listening to each other. From leader to teammates, teammates to leader, and among teammates, there should be consistency, clarity and courtesy. People should be able to disagree openly but with respect. Between the team and the public, responsiveness and openness is the key.

The difference between two equally talented teams is leadership. A good leader can bring a team success, provided values, the proper work ethic, and the vision of success. The Myth of the Head Table is the belief that on a team, one person is always in charge in every situation. Understand that in some situations; maybe another person would be best suited for leading the team. The Myth of the Round Table is the belief that everyone is equal, which is not true either. The person with greater skill, experience, and productivity in a given area is more important to the team in that area. Compensate where it is needed.

Make the decision to build a team and decide who among the team are worth developing. Gather the best team possible, pay the price to develop the team, do things together, delegate responsibility and authority and give credit for success.

Strategies to Follow When Starting a New Position

Any new situation can cause anxiety. The more you know and the more prepared you are about the situation, the easier it will be to adjust. When you begin a new position it is important to know as much as possible about the position and the ways that will best help you to ease into it. When you start a new position these strategies will help you ease into it

1. Prepare Yourself
Before you start your new job, get prepared and do your homework. Read everything you can, and speak to the smartest observers to understand the company's competitive positioning, assess its issues and opportunities, and evaluate the organizational culture and the quality of the management team.

If you'll be relocating, consider leaving your family at home for a while to immerse yourself in your new circumstances and to avoid the additional stress of having your spouse and/or children leave their friends and routines behind. At the same time, get in shape physically and emotionally. If you neglect your health, that's not going to do you, your company, or your family any good.

2. Set Realistic Expectations

Setting proper expectations is one of the most important things you can do when taking a new position, and it's also one of the most easily mishandled. A very common pitfall is to try to be a hero and make bold promises that can't be delivered or sustained.

In your earliest interactions with new employees, recognize that most will be listening through the lens of their own self-interest – "Will this new employee be good or bad for me?" So focus on answering the five questions: Who am I? What do I hope to accomplish? Where did I come from? Why am I here? And how do I hope to do it? Most important, don't be a know it all. Remember that you're the new kid on the block.

3. Establish a Productive Working Relationship with Your Manager

The best place to start establishing a productive relationship is to understand their motivations. This includes the straightforward goals of building shareholder value and achieving strategic, financial, and organizational objectives.

The savvy employee is also sensitive to the unstated motivations as well, such as the directors' concerns of minimizing their professional liability, maintaining their hard-earned reputations, and managing their busy calendars. In the case of managers, their greatest desire is generally to receive high-quality advice and loyalty, such that their own careers are enhanced.

4. Communicate!

Effective communication skills are the key to implementing just about everything. Successful communications starts with knowing your audience and establishing an emotional connection to your message. If you find yourself in a crisis, get the information out immediately and acknowledge the real challenges of the situation to establish credibility, which underlies all effective communications. Make sure not to come across as a know-it-all, which would turn people off and shut you down.

Burning Bridges All Fall Down

Here is a story which substantiates the motto don't burn your bridges. Too many inexperienced engineers are to quick to tell their supervisors what they really think. Even worse what they really think of them. Learn to be diplomatic. Don't burn your bridges.

"My name is Carol and as one of the Human Resources representatives at a large Engineering company I hear a lot of stories. This one takes the cake. It is the story of Wayne, a young Engineer we hired about 6 months ago. He was a little rough around the edges, meaning he had a temper, but he had all of the technical skills we were looking for so we hired him anyway. His manager here is someone with a lot of patience and we thought things would work out fine. Earlier this afternoon, he came into my office looking quite distraught. When I asked him what the problem was, he related this story to me."

"Before coming here Carol, I was working for Acme Engineering. I always did a good job technically, but I was having a difficult time getting along with my manager. It seemed like he was always on my back for one thing or another, things that seemed trivial and unreasonable to me anyway. One day, I just couldn't take it anymore and things got out of hand. After listening to one of his tirades, I threw all my papers at him, told him off good and stormed out of his office. When I got back to my cubicle, I hastily wrote out a seething resignation letter, put it in an interoffice envelope, addressed it to him and stormed out the front door of the building, never to come back again. I know I did the wrong thing that day, but that's how I came to interview here in the first place last year.

So last month, as you know Carol, we bought Acme Engineering and integrated them into our staff. After all the dust settled from people being transferred around, I now find myself working for this same man again. I just can't do it and I'm quitting as of now. Here's my resignation letter."

The moral of the story is when you change jobs always leave your old company on good terms with everyone. Don't burn any bridges. You never know when you might find yourself working for these same people again. Engineering communities are small and many employees move around within the same group of companies. I've been in Human Resources for many years and I've found myself working with some of the same people over and over again.

Client Relations

Not all engineers get to work directly with clients, which is especially true for new engineers. But eventually, during the course of your career you will have to interact with a client. The main point to good client relations is developing the ability to sense when the client is resisting. Unless directly countered, resistance tends to grow. Therefore, nip resistance in the bud, before it grows to an unmanageable proportion.

Resistance can take many forms. Here are just a few. The client may:
        Not agree with whatever you say
        Not like your credentials
        Be uncommunicative
        Confuse the subject
        Not tell you the truth about the project
        Interrupted the meeting
        Want quick solutions
        Use scapegoats
        Only want things done their way
        Ask for unreasonable guarantees

The issues underlying resistance are control and vulnerability. The client may view the problem as a pain or an embarrassment. You, on the other hand have to look at it as a way of helping the client get rid of the problem. To deal with resistance, first recognize the form it is taking. Then state to the client in a non blaming, neutral and diplomatic manner the kind of resistance you are experiencing. Doing this interrupts the client's pattern. Next, provide graceful egress for the client by tactfully asking for clarifications. Then listen.

For example, suppose a client is asking you many questions about the theoretical aspects of a product's operation. You realize that these questions are irrelevant. It seems that the client is trying to give you some sort of test to find out about your expertise. It makes you wonder if he is trying to diminish your credibility. You could counter this by saying in a nice voice, "You are asking a lot of questions about fundamental principles. Do you have any reservations about my ability to handle them? If not, I would like to proceed to the main issue". Nine times out of ten, the client will stop the interrogation and let you get on with the meeting.

Sometimes the only way to handle resistance is to be very assertive. Dealing with unrealistic customer expectations can be a challenge. A client may expect you to be familiar with a particular device that happens to be the only one of its kind in the country and that has been kept under close wraps. All you can do in this situation is inform them that it will take time to understand the problem in sufficient depth to produce useful results. Don't feel bad because you don't know the device. You have to get your own information in order to evaluate the situation.

Sometimes a client may withhold information that you need to complete the project. There are many possible motives for withholding information, but you have only limited visibly into the politics of the situation. If you cannot get the needed information through the back door, third party, or unofficial channels, you might be able to develop your own data, or find a way to make the information irrelevant.

If it agrees with your personality, learn how and when to tell jokes to create a light mood. Accumulate good jokes and anecdotes in a file and practice delivering them. Avoid shaggy dog stories and elaborate anecdotes that beg the listeners' patience. Consider your audience carefully. Laughter has an amazing effect, it releases tension. Dealing with clients is stressful enough for both parties. By allowing the element of humor to occasionally shine through, you will feel more relaxed and confident.

Conclusion

The many relationships in the engineering profession were discussed in this chapter. Working with clients was described and the major points of the relationship were outlined. Sample solutions of how to deal with resistance and other obstacles with clients were given.

A comprehensive list of successful team strategies were outlined to help you succeed while working in a team. Self questioning procedures enable you to understand what type of team player you are.

Meetings and Conferences

Meetings

Engineers are asked to attend all sorts of meetings within their company. As a Junior Engineer you will normally have only one regular meeting to attend, called by your immediate manager. It is a project status meeting. At the very beginning of your career, you won't be called upon to speak although after a short time, you and everyone else, in turn will report on the status of their portion of the project along with any special accomplishments or roadblocks they may have to completing their portion of the project on time and within budget.

These meetings are usually scheduled for the same day and time every week, biweekly or monthly. The frequency will depend on the complexity of the project you are working on and how often your manager feels he may need status from you. It is important to stay positive at these meetings and not to complain about other people on the team. Take responsibility for your accomplishments and failures alike. The team, and therefore the project is only as good as the weakest link. Don't be afraid to say that you don't know how to do something, or you don't know what the next step should be because your team members are there to help you. Speak up if a technical aspect of your project has you stumped.

From time to time, your supervisor will also call you to his or her office for a one-on-one. At these meetings, you will be able to talk in detail about the special issues of what you are working on and you will both be more free to speak your minds especially concerning other employees, but remember — tread lightly.

As you progress through the ranks of your company, you may be invited to strategy meetings where they talk about long term goals and progress as well as plans for new products. In addition, one or several Engineers on your team will be invited to Quality meetings, Manufacturing meetings, and Marketing meetings. If you get invited to any of these meetings you can consider yourself to be moving up in the minds of the other employees.

Take time to prepare for these meetings. Most companies expect you to use visual aids like Microsoft PowerPoint presentation software. You can prepare your presentation on the computer and print copies to distribute to the attendees when it is your turn to speak. If there is an overhead projector available you may print transparencies and if there is a computer projector available, you may bring your laptop computer and just plug it in when it's your turn. After these types of meetings, your manager, if he was not in attendance will want to be debriefed as to what was discussed.

If you become a manager yourself, you will find that most of every day is spent attending meetings. You'll find yourself working long into the evening just getting your own work done as well as preparing for the next day's meetings.

Your status within the company can be determined by the number and types of meetings you are invited to attend. The more meetings you're invited to, the more valued your input is to the rest of the company. Keep a good record of your calendar, you don't want to be late for meetings. Most meetings are planned using software such as Meeting Maker. They come with reminders that will flash on your computer screen several minutes before you are due to attend. It's important that you're not the last one there! Keep good notes at meetings so you will be prepared if changes need to be made to your portion of the project and for debriefings with your manager which can sometimes be as much as a week after the actual meeting occurred.

Most companies have quarterly all-hands meetings where the President or CEO as well as some of the Vice Presidents will speak about the health of the company in general as well as their visions for the future. These meetings tend to have a very relaxed and gung ho atmosphere and provide an opportunity for you to mingle with other employees who you wouldn't normally come into contact with.

In general, meetings are a means of showing your face. The more people within the company who know who you are, the better your chances for promotions or transfers are. If a more senior opening in another department presents itself and that manager doesn't know who you are, there is no way you will be considered for the position. There is nothing more satisfying than having a manager from another department ask you if you are interested in joining their team, at a higher salary and with more responsibilities, of course.

Conferences

Engineers have the opportunity to attend different types of conferences, most of which are held outside their company's facilities, or even in other cities around the country. Technical conferences are the most common.

With technology changing so rapidly, it is important for your company to have you attend as many as they can afford to send you to without jeopardizing your project's progress. Companies are keen to have you learn new things within your field and will usually pay for you to go to technical conferences regularly. Many companies have a policy of sending their engineers to at least one out of town technical conference every quarter. You may not be chosen to go to every one, but when you are, when you return, be prepared to debrief your engineering team on the technology that was discussed.

Other conferences you attend are more on a voluntary basis. You may find yourself going to San Francisco, Las Vegas or New York City conventions for the purpose of gathering literature from your competitors and assessing their products. When you return from these types of conferences, you will also be asked to make a presentation of your findings to various groups within your company. You may also find yourself attending sales conferences, attended by your company's field sales force for the purpose of making a technical presentation for the project or product you are working on.

Most Engineers have strong feelings about meetings in general. Many are intimidated by the prospect of having to stand up and make presentations to their peers and supervisors. Others think of meetings as an opportunity to socialize and meet fellow employees from other departments, while still others feel they are a waste of time and would rather be back at their desks or in the lab working on purely technical matters. Progressive companies and companies that are doing well will have food brought in at meetings which for the average engineer is a big plus. Meetings may occur any time of day. There are breakfast meetings, lunch meetings as well as after hours meetings. They are all important and can lead you to getting ahead in the company.

Sales Meeting

Even though the majority of engineering is not sales related, sometime during the course of your career you will probably find yourself in some type of sales meeting. Engineers tend to be uncomfortable with sales people but there is no substitute for the sales meeting. Alternatives are generally a waste of time. You must learn to develop sales attitudes and abilities. Sales abilities can be developed with a modest amount of effort, at least to the point where you are not making major mistakes. The following techniques will give you a jump start.

 Project a confident image
 Gain rapport
 Establish a need

Show how you can satisfy the need
Elicit the client's objections and concerns
Lead the client gently to agreement

In order to project a confident image you have to believe in yourself. Your inner attitude shows through. Once you have this sense of self esteem you do not need to rely on tricks or high pressure tactics to get contracts, you can simply use your ability to solve the client's problem.

The goal of rapport is to break down the barriers that prevent straightforward communication. You will know you are successful when you establish a dialogue that is balanced and flowing. Become familiar with the client's business and operation beforehand. Make sure your appearance fits in with the client's corporate culture. Then, use your conversational skills to establish points of common concern.

First talk about neutral subjects gradually working into more personal subjects that indicate a common interest or viewpoint. Avoid controversial topics such as politics, sex and religion. Aim for popular subjects such as sports, the company's background, or highlights particular to the client's location. When you attain rapport, the quality of conversation becomes 'our' interests rather then his and mine.

Once you have established rapport you are ready to establish the client's needs. In this phase of the meeting concentrate on the customer's problem. If you start to think of your own problems you'll distract yourself and lose the ability to think. The conversation should center on the client's problem. Ask open ended questions, those than cannot be answered by a simple yes or no answer. Make every effort to be nonjudgmental in asking questions. If you don't keep your prejudices to yourself you run the risk of self sabotage. You don't want the dialog to go like this.

You: "You're not going to use the Star Trek model 6 recorders, are you? It's a loser!" Be able to speak intelligently and back up your points. That allows you to say whatever you want with impunity.

Client: "Well, our president insists it's the best unit on the market now and I agree with him"

Of course you must listen to the answers. You can help the client feel comfortable during the fact finding stage by being a good listener. Listening is an art. You cannot talk people into buying but you can listen them into it.

Here are some listening skills:

    Be open minded

    Give the client your undivided attention

    Politely ask for clarification on points you don't understand

    Listen for emphasized words

    Listen and look at body language

To show how you can satisfy the client's needs, convince them that you are not selling your product but fulfilling that need. You are not offering technical services but solving your client's critical problems. Let them reach the conclusion that it's your product that they want.

Engineers are notorious for trying to sell a product based on its features, features that they themselves may have designed into the product. It takes a long time for technical people to realize that customers do not buy a product or service based on its features but rather on the perceived benefits. What does this mean to you? It means you cannot try to sell the client on your expertise or technical capabilities. Instead focus on the client's problem and explore the ways you can solve it.

Sales meetings with new clients are almost always challenging. To rise to the challenge you must adequately prepare and think quickly on your feet. You should have asked the following questions before the meeting:

    What is the need, both technically and project wide?

    What is the scope of the work?

    When does the client need the work done?

    Is your company the only company responding to the client needs?

Before the meeting, look up recent articles on both the technical subject and the project. Bring reports, photographs and drawings that substantiate your background in your technical area.

As you get further into the sales meeting, think of good reasons why your service will offer more value than that of your competitors. Show the client how your service is:

More reliable

Faster

Better at technical support

More cost efficient

More convenient

More credible

Spell out the benefits. It is a mistake to assume that your clients can figure these things out for themselves.

In order to draw out the client's objections and concerns, ask to understand them in greater depth. Don't defend yourself against the objections, explore the objections. In the process you will learn enough detail to structure a response that emphasizes your strengths. For example, "I can understand why maintaining low hourly labor rates are important for you. However, I can show you how many of my clients save money on the overall project by using my computer program"

Don't react to objections by arguing. Make every effort to prevent objections from turning into verbal boxing matches and never tell a client he is dead wrong. Maintain a friendly attitude, repeat the objection to the client in your own words and ask them to be even more specific. "In what respect do you feel it is inappropriate, or I can see how you can feel that way, many others have felt that way too". The next step is an important one, by saying something like "But, I can give you three good reasons why it is to your advantage to consider another viewpoint for a moment." Then explain how these benefits have greater value.

Now that you have shown the client how you can solve his problem it's time to wrap things up. By this time you will know whether the client is interested. Closing the sale takes the client from interested to commitment. Toward the end of the meeting, begin making hints about including certain items in your proposal. If interest has been good close with "I'm very enthusiastic about working with you on this project. What is our next step? What paper work do we need to make this official?"

Some clients say they want to think about it. In that case offer to go back to your office and consider the problem more. Often, the client may be unsure because he is considering other options at the same time. Agreeing to wait is fine as long as you can call the client later and monitor the status. If the client hesitates because he doesn't know when the project will start, convince him that there is no risk in writing you a contract now. Offer assurances that you won't start work on the contract until the client turns it on. If the client is unsure of your ability to handle the project offer to send the telephone number of references who can attest to your competence and capabilities. Don't use high pressure tactics to close the sale. Convince them that it's in their best interest.

Guidelines For a Sales Meeting

Some basic guidelines that will help you have a successful sale meeting are:
　　Listen carefully
　　Try to determine the competitive basis of your bid
　　Be in tune to the client's problem but don't feel you have to solve it today.
　　Discuss your past and present projects.
　　Never bluff
　　Think positively
　　Conclude the meeting with an agreement for action

Visual Aids

Many people are visual thinkers. Sometimes an excess of words can turn them off. By using a picture show, you get a chance to show them more of what you can do. If you try to do this verbally you run the risk of running out of time before you have said enough to convince them.

Make a portfolio of your previous projects. It could be leather bound; a three ring binder that contains a number of transparent polystyrene sheet protectors containing a sketch or photo with a few words describing a particular project. Use color in as many of the sketches as you can. Include large photos depicting yourself with the equipment or clients. Arrange them so the idea of the project can be grasped in a single glance. The main purpose is to give a quick visual impression. Make sure everything is neat and professional looking.

The portfolio concept can be extended many ways. You could bring in models of your work or actual hardware, if they are small enough. Or, using a laptop computer you can even create an electronic portfolio.

Conclusion

Meetings and conferences are outlined in this chapter and you now have a better picture of what happens at them and what is expected of you. Group, one on one, strategy and quarterly meetings are some of the types that are discussed in detail so you have hints on how to prepare and shine during any type of meeting. You have discovered that the more meetings you are invited too the more valued your input is. You now know meetings are an important means of getting known in the company and how they help you achieve promotions or transfers.

The difference between meetings and conferences are described. You are now knowledgeable on the most common type of conference, the technical conference, and why it is so important. This chapter also described how conferences are usually a voluntary situation and are held in major cities. You now know why your company would send you to a conference and what they expect from you in return. Sales meetings are discussed giving you helpful hints on how to succeed during a sales meeting.

This chapter brings up an important reason for meetings; they are tools to get promotions, transfers and becoming known within the company. You now know how to use them to your advantage.

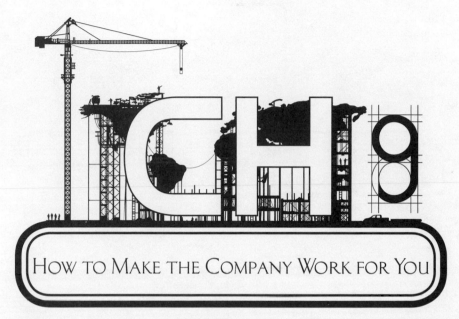

HOW TO MAKE THE COMPANY WORK FOR YOU

How Companies are Organized

Engineering companies are organized very formally and there are employees who spend the better part of their day drawing organizational charts, usually referred to as 'org charts'. For a typical Computer or Electronics company, an org chart may look like this.

In a small company, an employee may wear several hats. The VP of Sales and the VP of Marketing may be the same person. The Director of Hardware and the Director of Software may also be the same person. As an Engineer, you are on the bottom rung of the ladder among the salaried employees. Below you are the hourly employees, like technicians and administrative people.

As an Engineer, you usually have the choice over the years of becoming a specialist or a generalist. Many companies will move you around periodically and you will become a generalist by default. If you are really

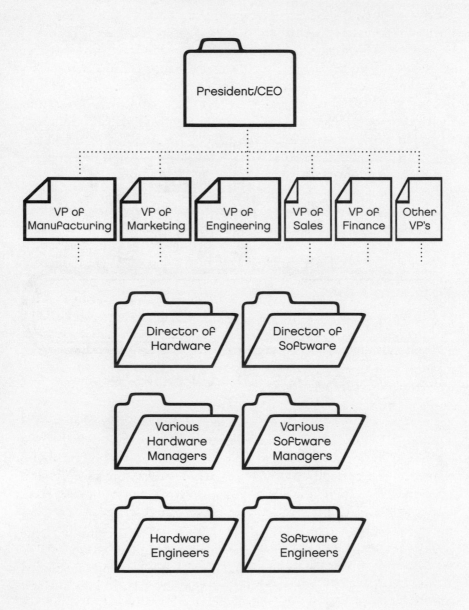

good at one particular aspect of the business you will usually get assignments within your expertise and become an expert in that field. If you enjoy the hands-on aspect of engineering and want to remain an engineer throughout your career, you should strive to become a specialist. If you are more interested in getting into management, you should strive to become a generalist. Specialists rarely become managers. Because of their expertise, they become indispensable at the jobs they do. Although they are rarely promoted into management, because of their expertise, they may earn salaries up to $150,000 in their field. If you find that you are management material, then salary-wise, the sky is the limit but with each promotion, the choices become fewer and fewer. Keep in mind that there may be 100 engineers reporting 10 engineering managers, who all report to only one Vice President!

Generally, there are no unions for Engineers so your raises and promotions are due to your technical expertise as well as your ability to function in a team environment. As a new graduate, getting used to working in a team environment is most important. Projects are usually large enough that you will be working within a group, with you being responsible for only one aspect of the project. It is important to communicate amongst yourselves because the success of the project is only as good as the various aspects come together.

As in any other business, when you are first starting off, don't be thought of as a know-it-all. Keep quiet at meetings and do your work in a timely manner. Typically, you will spend your day in front of computer working on the technical aspects of your project. As you become more senior, you will be asked to attend more meetings, get involved in manpower forecasting as well as writing projects in concert with Technical Writers.

If you were in the middle of your graduating class in college, your first job may be as a Manufacturing Engineer or a Test Engineer within the Manufacturing Department. Mostly the top tier of new graduates get the opportunity to go directly into design and development of new products, but don't despair, Test and Manufacturing Engineers do get transferred into design and development after demonstrating superior abilities in whatever they are doing.

Most large companies like to transfer a certain percentage of Manufacturing and Test Engineers into Design and Development every year. It's a sign of wanting to promote from within. Employee retention is important for most companies because the cost of training is so high so they make an effort to make these types of transfers whenever possible.

As an entry level engineer you will usually do routine work for three to five years. You may receive formal classroom or seminar type training but mostly they expect you to know how to do the job.. With experience you will begin to take on more difficult tasks and then allowed to be more independent in developing designs and solving problems.

You may advance to become a technical specialist or a manager. You may eventually join engineering management or enter other managerial or sales positions.

Some engineering oriented companies have created parallel paths that allow engineers to receive substantial promotions and raises without moving into management.

If you do choose a management track you may begin as a project leader supervising three to five engineers. A group manager supervisees the activities of three to five project managers. A department manger is often responsible for 10 to 50 engineers and a budget of more than one million dollars.

Companies Responsibilities

As a new engineer there is no promise that your employer will be committed to optimizing your advancement. Don't be confused by unrealistic expectations fostered by engineering schools and corporate personnel offices. The plain fact is that a corporation does not owe its employees much. You and the other corporate employees are hired workers. Don't allow your ego to not accept this. Don't develop blinders to the business realities surrounding your employment. The first responsibility of a corporation is to its stock holders.

You must learn to develop the perspective that will allow you to grow. You can become successful in engineering if time and energy are expanded to build up your personal confidence and enthusiasm, despite the shortcomings of the field. The price of not becoming a victim is hard work and learning to see the positive goals and not the ocean of negatives surrounding it.

Thoughtful consideration of your own interests, values, and goals is essential but it will not be enough. You must also consider the goals, values, and culture of the organization in which you decide to work. Is the organization a good fit for you? Keep in mind that your employer is only interested in your growth and development to the extent that it is beneficial to the company's organizational goals. Ultimately you are responsible for your own professional growth and development.

Positive Thinking

As an engineer, you have a good chance of working in an environment that is not conducive to developing attitudes of professional independence or business self identity. You may, for many years be exposed to people marching in line and become hypnotized into "can't do" thinking. They focus on reasons why things won't work or how cost, time and expediency make change unfeasible. After prolonged exposure to this environment you may become absorbed in this organization inertial. But don't look to such people for advice or encouragement as they do not value the very qualities you will need to survive such as creativity, efficiency and professionalism.

Be aware of your own negative thoughts, they can keep you in a rut. Don't lose interest in your career and be careful not to give up. It is a mistake to assume that a corporation or any organization representing the interests of its owners will love you for better or worse. You should not assume that some business entity will look after your own long term welfare. You should continuously monitor whether your career goals are being achieved within that company. Other opportunities available to you may lead you closer to your goals.

Company Rewards

Working for a large company is an advantage if you are "average" and is a blessing if you are "below average". Many engineers of moderate talents make it to the top of their companies simply by putting in their years. Talent and technical performance may not be the sole criteria of advancement. Other factors such as effort, loyalty, and the ability to deal with people might carry more weight.

Other factors such as effort, loyalty, and the ability to deal with people can carry more weight. Nevertheless, if you are aiming for a top position in a large company's engineering section, it is worth considering the small numerical probability of your rapid advancement.

Your Responsibilities

As a new engineer you will have to be more versatile and sensitive to the changing needs and expectations of industry. You also have to think about your career in different terms than the engineers who have gone before you. As in the past, creativity, problem solving, and technical expertise will be important. However in today's engineering market you will have to have more. You will need to demonstrate teamwork, leadership, multicultural awareness and entrepreneurial spirit.

The Right Track

As you examine the professional and educational experiences of the senior management in various businesses and industries you will begin to notice that in some industries the way to the top is through the technical track. In others you need to take the management track. If you have your goals set on being part of an executive team, you need to look at industries which offer you the opportunity to go as far as you can within the track that best fits you. Therefore it is important to know which functional area or areas will provide you with these opportunities for personal and professional growth.

The functional areas in the technical track are research and development, production, and technical professional services. The functional areas in the management track are marketing and sales, information systems processing, accounting and finance and administration.

Research engineers are engaged in systematic and critical investigations leading to the acquisition of knowledge for a specific application. Engineers in production are central to the mission of the company or the industry as a whole. They are in the area where products are taken from the research and development stage to finished marketable products. Technical professional services engineers perform functions ranging from design and testing to feasibility studies and consulting.

In many companies marketing and sales is one of the better paths to corporate management for engineers as it not only requires technical expertise but also provides a broad overview for the customer base, research and development, production and distribution. Engineers involved in information systems, process, analyze, design, develop, and implement those systems that are the backbone of the company. In accounting and finance, engineers are involved in financial analysis, management consulting, operations research, strategic planning and actuarial work. Engineers in administration are in non technical aspects of business.

Small and Midsized Companies

You may want to look into working for a small or midsized company as there are a number of opportunities for new engineers in these companies. If you have an entrepreneurial sprit and want to find exciting opportunities this might be for you. When managed correctly these size companies can become the conglomerates of the future. However it is still important to research them well because if they do not make it, you could be out of a job faster than if you were with a large organization.

Here are some suggestions. Research the owner's background and reputation through local professional organizations. Research the technology and or the processes. Does the owner hold the patent for the technology? Are there any financial records available for review? It pays to be caution but the challenge and excitement of being part of a growing company is something that should be seriously considered.

Get Company Information

It is wise to get as much information as you can about companies that you are interested in. Their annual report is an extremely helpful piece of information. Begin by reading the chief executive officer's letter to shareholders. It will address the challenges and the achievements of the company over the past year. It will also describe a plan of action for the future. The letter will also indicate how employees will be affected by the plan.

Next look at the gross sales and the expenditures of the company. Are there any areas that appear to be larger than expected, or significantly higher than the previous year? This may be a good source of questions about what was going on within the company to require such a cost. Acquisitions and new product lines may also be identified in reading the annual report. It is a wealth of information and could help you in gaining an edge in the interview process.

Conclusion

Every company has an organizational structure and this chapter showed you a typical organizational chart for an electronics company. Different types of companies such as small, medium and large were discussed and the different roles an engineer may have in them.

Choices of becoming a specialist or moving into management are some of the aspects described. The non existence of unions and how raises and promotions are given is outlined in this chapter. What you probably will do on you first job is described and hints on how to act for success are discussed. Steps in the advancement process are given and you now have a better idea of how advancement takes place and what to do to achieve it.

The realization of you being responsible for your own success in the company is stated. The plain fact that a corporation does not owe you very much is recognized. You now know you must consider the goals, values, and culture of the organization that you work for as well as you own interests, values and goals. Keeping an optimistic attitude is also essential for your success and the threat of negativism is discussed.

Your personal and professional growth in the company and how to achieve your goals are explored. Also the different tracks you can take such as technical verses management track were defined and you now have a better understanding of which is for you.

THE TEN COMMANDMENTS OF ENGINEERING

❶   Keep learning

❷   Pick the right college and major

❸   Pick the right field

❹   Never talk badly about your coworkers or manager

❺   Make meetings work for you

❻   Make the company work for you

❼   Be a team player

❽   Get the right engineering position for you

❾   Get the right license and degree

❿   Always perform your work in a legal and ethical manner

1. Keep Learning

Engineers usually prefer learning through lots of action and hands on activities. They may be experimenters or tinkerers. They most likely enjoy competitions, challenges, and taking risks when learning. They are practical and want to get the straight facts on any subject. They can become bored and disengaged when presented with abstract theory unless a logical connection can be drawn between the theory and practice.

Not only will you need a bachelor degree but most firms also require a master's degree, so stay in school as long as you can. It is also important for engineers to continue their education throughout their careers because much of their value to an employer depends on knowledge of the latest technology. Because technology changes so rapidly, most companies offer paid training for their engineering staff, and virtually all offer reimbursement for taking college courses within your field.

The need for continuing education is a common denominator in the engineering profession. Larger companies may offer formal classroom or seminar training. A rule of thumb for continuing education is 5% of your working time should be dedicated to learning new or advanced topics in your field or in a new emerging field.
Refer to Chapters 1 and 4

2. Pick the Right College and Major

The first thing is to make sure the college or university you attend has an ABET accredited engineering program. You need to be careful because some departments or programs within a college or university can be ABET accredited while others are not.

You will need a BS degree for any type of engineering you decide to go into. A master's degree is preferred and even a PhD may be necessary to obtain some positions in research, education and administration in most engineering fields. Most engineering degrees are granted in electrical, electronics, mechanical or civil engineering.

College graduates with a degree in the physical sciences or mathematics occasionally may qualify for some engineering jobs. However, engineers trained in one branch may work in related branches. Many aerospace engineers have training in mechanical engineering. This flexibility allows employers to meet staffing needs in new technologies and specialties in which engineers may be in short supply. It also allows you to shift fields with better employment prospects.

A graduate degree is a prerequisite for becoming a university professor or researcher. New graduates with MS degrees also command higher starting salaries. A substantial number of engineering graduates combine their engineering degree with a master's degree in business administration (MBA). The engineering/MBA combination is a powerful one for engineers who expect to move into management or to start their own business. Engineers whose work may affect the life, health or safety of public must be registered according to regulation in all 50 states and the District of Columbia. You must have received a degree from an accredited engineering program and have four years of experience. You must also pass a written examination. Many engineers also become certified. The doctorate degree (PhD) is obligatory for engineers who expect to teach at the university level.

When you make plans to attend college and study engineering you should consider several factors: the type of engineering programs they offer, tje quality of the school, employment possibilities and your personal preferences in living and learning conditions.

In the United States there are more than 300 colleges and universities where engineering programs have been approved by the Accreditation Board for Engineering (ABET) and Technology. It is wise and will help you with licensing if you attend a ABET accredited engineering program. Make sure all the departments or programs within the college are ABET accredited. Although admissions requirements may vary, most require a solid background in mathematics and science. ABET accreditation is based on an examination of an engineering program's student achievement, program improvement over time, faculty, curricular content, facilities, and institutional commitment.

Although most educational institutions offer programs in the major branches of engineering, only a few offer programs in the smaller specialties. Also, programs of the same title may vary in content from school to school. For example some programs emphasize industrial practices preparing students for a job in industry, while other are more theoretical and are designed to prepare students for graduate work. Therefore you should investigate curriculum and check accreditations carefully before selecting a college.

An engineering education will teach you quantitative reasoning, problem solving and design. Your overall college education will teach you critical thinking, writing, and speaking, all of which can be successfully applied to a number of different jobs in the field of engineering. But it still remains up to you to choose a job and to learn how to use the benefits of your education in a way your employer will appreciate.

An MBA or technical master's degree enhances your opportunities for promotion. A doctoral degree is essential for university teaching or supervisory research positions, but may not help in an industrial setting. When you get to this point you may decide to start your own consulting firm. Continued advancement, raises and increased responsibility are not automatic but depend on sustained demonstration of technical expertise and leadership abilities. While the choice of school is important for personal reasons, most engineers will find that individual performance is more important than the school as a measure of success.
Refer to Chapter 4

3. Pick the Right Field of Engineering
Since there are many types of engineering fields, you should narrow it down to the field that you are most interested in and that you are most compatible with. First, what are you best at doing and what do you like doing? Find out what you are best at and interested in by volunteering, examining your hobbies, doing summer internships etc.

All engineers engage in one of five areas of work: research, development, application, management, and maintenance. Engineers who work in research are responsible for investigating new materials, processes, or principles for practical applications of ideas and materials. Engineers who work in development use the research results to determine how best to apply them to their practical functions. Application engineers produce the actual materials, machines, and methods designed by research and development engineers. Management and maintenance engineers keep the developed idea working and make improvements and adjustments. There are six areas of engineering that form the traditional areas of the profession. These areas are chemical, civil, electrical, industrial, materials science and mechanical engineering. Preparation in any one of these areas will provide a solid foundation for a wide range of specialties.

Over the years, each traditional branch of engineering has developed into focused specialties. Today some of these specialties have become engineering professions in their own rights such as aerospace, agricultural, automotive, biomedical, computer, environmental, manufacturing, mining, traffic, metallurgy, nuclear, optics, petroleum, plastics, packaging, quality control, and robotics.

Depending on what type of engineering you want to go into, the demand will vary. The general overall growth of engineering is expected to grow more slowly than the average for all occupations over the 2006 to 2012 period. Engineers tend to be concentrated in slow growing manufacturing industries, a factor which tends to hold down their employment growth.

Also many employers are increasing their use of engineering services performed in other countries. Despite this, job opportunities for engineers are expected to be good because the number of engineering graduates should be in rough balance with the number of job openings over this same period. Some areas will see short term growth and still others will bloom. Projections range from a decline in employment of mining and geological engineers, petroleum engineers, and nuclear engineers to much faster than average growth among biomedical and environmental engineers.

A career in engineering provides you with many opportunities for work settings. Most engineers work in office buildings, laboratories, or industrial plants. Others may spend time outdoors at construction and production sites where they monitor or direct operations or solve onsite problems. Some engineers travel extensively to plants or worksites. There is a good chance you will be able to find a position in any setting that suits your needs.

Any position can be stressful but depending on what type of engineering you go into the stress level can be more then the average job. At times, deadlines or design standards may bring extra pressure to a job, sometimes requiring engineers to work longer stressful hours. Engineering is definitely not a 9 to 5 job and tends to be project oriented meaning that your work on a particular project must be completed by a certain deadline regardless of the hours you need to put in to get it done.

If you work for an engineering consulting firm, you will perform engineering tasks for other companies or organizations and when your job is done you move on to a new project with another company or organization. You work on numerous projects with different types of organizations and people. Some engineers pursue consulting careers early to help them decide where they ultimately want to work.

The government career path can be an excellent choice. The federal, state, and local governments can be excellent employers. NASA employs many types of engineers in the space program because of the great diversity and creativity of its projects. The US Environmental Protection Agency and the U.S. Army Corps of Engineers employ many civil and environmental engineers; there are also numerous career paths for these engineers in state departments of transportation and environmental protection agencies throughout the country.

The Food and Drug Administration is a major employer of biomedical engineers while the U.S. Department of Defense continues to employ a wide variety of engineers in both civilian and enlisted positions for its agencies and installations throughout the world.

Most engineers who pursue an academic path have received either a master's or PhD. degrees and teach in colleges and universities. Some engineers decide to obtain state teaching certificates after completing their bachelor's degree and become science or math teachers in middle schools or high schools.

The internet career path has opened new possibilities for engineers. There are opportunities for engineers with companies that are associated with the internet. There are also internet career possibilities with more traditional companies. Companies have realized that to be successful they need to use the Internet to get their products to their customers more quickly and more cost effectively.

About half of all engineers work for companies that design and manufacture electronics, scientific instruments, automobiles, aircraft, chemicals and petroleum products. Biomedical engineers work at universities that are affiliated with hospitals or medical complexes. The remaining group is divided between public utilities and federal, state and local government departments and agencies, from NASA to state highway departments.

If you have your goals set on being part of an executive team someday, you need to look at industries which offer you the opportunity to go as far as you can in the track that best fits you. Therefore it is important to know which functional area or areas will provide you with these opportunities for personal and professional growth.

The functional areas in the technical track are research and development, production, and technical professional services. The functional areas in the management track are marketing and sales, information systems processing, accounting and finance and administration.
Refer to Chapters 2, 3, and 5

4. Never talk badly about your coworkers.
Many relationships exist in the engineering profession. You must learn what they are and how to behave for best results. To many, inexperienced engineers are too quick to tell their supervisors what they really think of the department or company. Even worse, what they really think of them. Learn to be diplomatic. Don't burn your bridges. When leaving a job never talk badly of your coworkers. You never know when you may find yourself working for them in the future. This is only common sense
Refer to Chapters 7, 8, 9

5. Make Meetings Work for You
Engineers are asked to attend all sorts of meetings within their company. As a Junior Engineer you will normally have only one regular meeting to attend, called by your immediate manager. It is a project status meeting. At the very beginning of your career, you won't be called upon to speak

although after a short time, you and everyone else, in turn will report on the status of their portion of the project along with any special accomplishments or roadblocks they may have in completing their portion of the project on time and within budget.

These meetings are usually scheduled for the same day and time every week, biweekly or monthly. The frequency will depend on the complexity of the project you are working on and how often your manager feels he may need status from you. It is important to stay positive at these meetings and not to complain about other people on the team. Take responsibility for your accomplishments and failures alike. The team, and therefore the project is only as good as the weakest link. Don't be afraid to say that you don't know how to do something, or you don't know what the next step should be because your team members are there to help you. Speak up if a technical aspect of your project has you stumped.

From time to time, your supervisor will also call you to his or her office for one-on-one. At these meetings, you will be able to talk in detail about the special issues of what you are working on and you will both be more free to speak your minds especially concerning other employees, but remember – tread lightly.

As you progress through the ranks with your company, you may be invited to strategy meetings where they talk about long term goals and progress as well as plans for new products. In addition, one or several Engineers on your team will be invited to quality meetings, manufacturing meetings, and marketing meetings. If you get invited to any of these meetings you can consider yourself to be moving up in the minds of the other employees.

Take time to prepare well for these meetings. Most companies expect you to use visual aids like Microsoft PowerPoint presentation software. You can prepare your presentation on the computer and print copies to distribute to the attendees when it is your turn to speak. If there is an overhead projector available you may print transparencies and if there is a computer projector available, you may bring your notebook and just plug it in when it's your turn. After these types of meetings, your manager, if he was not in attendance will want to be debriefed as to what was discussed.

If you become a Manager yourself, you will find that most of every day is spent attending meetings. You'll find yourself working long into the evening just getting your own work done as well as preparing for the next day's meetings.

Your status within the company can be determined by the number and types of meetings you are invited to attend. The more meetings you're invited to, the more valued your input is to the rest of the company. Keep a good record of your calendar. You don't want to be late for meetings. Most meetings are planned using software such as Meeting Maker. They come with reminders that will flash on your computer screen several minutes before you are due to attend. It's important that you're not the last one there! Keep good notes at meetings so you will be prepared if changes need to be made to your portion of the project and for debriefings with your manager which can sometimes be up to a week after the actual meeting occurred.

Most companies have quarterly all-hands meetings where the President or CEO as well as some of the Vice Presidents will speak about the health of the company in general as well as their visions for the future. These meetings tend to have a very relaxed if not gung ho atmosphere and provide an opportunity for you to mingle with other employees who you wouldn't normally come into contact with.

In general, meetings are a means of showing your face. The more people within the company who know who you are, the better your chances for promotions or transfers are. If a more senior opening in another department presents itself, if that manager doesn't know who you are, there is no way you will be considered for the position. There is nothing more satisfying than having a manager from another department ask you if you are interested in joining their team, at a higher salary and with more responsibilities, of course.
Refer to Chapter 8,

6. Make the Company Work for You
Typically, new graduates who are not in the top third of their graduating class begin their careers as test or manufacturing engineers. Typically, you are given the title of Junior Engineer. You often begin working un-

der the supervision of more experienced engineers. Many participate in special training programs designed to orient them to company processes, procedures, polices and products. This allows the company to determine where you best fulfill its needs. After this training period you often rotate through different positions to get an all around experience in working for the company. You are usually only allowed to do actual design and development work after the training process is complete. A new graduate with an MS degree may immediately begin doing original design work.

In addition to salaries, most private companies and many public companies offer stock options and have yearly bonus plans. Stock options alone create more millionaires among engineers than anything else.

As an Engineer, you usually have the choice over the years of becoming a specialist or a generalist. Many companies will move you around periodically and you will become a generalist by default. If you are really good at one particular aspect of the business you will usually get assignments within your expertise and become an expert in that field. If you enjoy the hands-on aspect of engineering and want to remain an engineer throughout your career, you should strive to become a specialist.

If you are more interested in getting into management, you should strive to become a generalist. Specialists rarely become managers. Because of their expertise, they become indispensable at the jobs they do. Although they are rarely promoted into management, because of their expertise, they may earn salaries up to $150,000 in their field. If you find that you are management material, then salary-wise, the sky is the limit but with each promotion, the choices become fewer and fewer. Keep in mind that there may be 10 engineering managers, but they all report to only one Vice President!

There are no unions for Engineers so your raises and promotions are due to your technical expertise as well as their ability to function in a team environment.

Most large companies like to transfer a certain percentage of manufacturing and test engineers into design and development every year. It's a sign of wanting to promote from within. Employee retention is important for most companies because the cost of training is so high they make an effort to make these types of transfers whenever possible.

As a new engineer there is no promise that the employer will be committed to optimizing your advancement. Don't be confused by unrealistic expectations fostered by engineering schools and corporate personnel offices. The plain fact is that a corporation does not owe its employees much. You and the other corporate employees are hired workers. Don't allow your ego to not accept this. Don't develop blinders to the business realities surrounding your employment.

You may want to look into working for a small or midsized company as there are a number of opportunities for new engineers in these companies. If you have an entrepreneurial sprit and want to find exciting opportunities this may be for you. When managed correctly these size companies can become the conglomerates of the future. However it is still important to research them well because if they do not make it you could be out of a job faster than if you were with a large organization.

 Research the owner's background and reputation through local professional organizations. Research the technology and/or the process. Does the owner hold the patent for the technology? Are there any financial records available for review? It pays to be caution but the challenge and excitement of being part of a growing company is something that should be seriously considered.

Most engineers are employees working for large companies. Some engineers become private consultants. This is usually a self employed career. Some Engineers ultimately become managers while other may have opportunities in marketing or sales.

It is wise to get as much information about a company that you are interested in. The annual report is an extremely helpful piece of information. Refer to Chapters 5 and 9

7. Be a Team Player
Working in a team is an important part of being an engineer. You are more likely to work as a team then on your own. To achieve great things,

you need a team. Building a winning team requires understanding of certain principles.

Whatever your goal or project, you need to add value and invest in your team so the end product benefits from more ideas, energy, resources, and perspectives. First people who try to achieve great things by themselves sometimes do so because of the size of their ego, their level of insecurity, or simple naiveté and temperament. Remember one is too small a number to achieve greatness. There is power in numbers. You will be more successful in a team than by yourself.

The team's goal should be more important than your role within it. Members of the team must be willing to subordinate their roles and personal agendas to support the team vision. By seeing the big picture, effectively communicating the vision to the team, providing the needed resources, and hiring the right players, you can create a more unified team.

All team members have a place where they add the most value. Especially, when the right team member is in the right place, everyone benefits. Refer to Chapters 1 and, 7

8. Get The Right Engineering Position For you.

First, in order to get the right position the main thing to do is network, network, and network. That is the key. Networking will help you become more knowledgeable about all the employment opportunities available during your job search. Everyone you know is part of your network, even if they are not engineers. They may be able to introduce you to someone they know who works in the engineering field you're targeting.

Most engineers obtain their first position through company recruiters sent to college campuses. If you find yourself face to face with a representative of the company's human resources department you will want to emphasize your interpersonal skills and willingness to learn rather than engineering lingo. If your interpersonal skills could use a little fine tuning have someone in the placement office or a friend videotape you in a mock interview.

You can also find employment through a summer position or work study

arrangement. Cooperative experience in college lends a big hand not only because of the exposure to the actual field you're interested in but also the possibilities of after college contacts. Keeping in touch with fellow engineers leads to even more contacts. You should respond to advertisements in professional journals or newspapers. The internet now offers multiple opportunities to job seekers. Join a professional engineering organization and use their employment services. Also look into employment agencies and headhunters that deal with engineers. Interview with everyone for experience and possible connections.
Refer to Chapters 1, 2, 5, and 9

9. Get The Right License and Degree.
All 50 states and the District of Columbia require licensure for engineers who offer their services directly to the public. Engineers who are licensed are called Professional Engineers (PE). This licensure generally requires a degree from an ABET accredited engineering program, passing the Fundamentals of Engineering (FE) exam formerly called the Engineering In Training (EIT) exam, four years of relevant work experience, and successful completion of a State Principal of Practice of Engineering (PE) examination.

Recent graduates can start the licensing process by taking the examination in two stages. The initial Fundamentals of Engineering (FE) examination can be taken upon graduation. Engineers who pass this examination commonly are called Engineers in Training (EIT) or Engineer Interns (EI). After acquiring suitable work experience, EITs can take the second examination, the Principles and Practice of Engineering exam. Several States have imposed mandatory continuing education requirements for re- licensing. Most States recognize licensure from other States provided that the manner in which the initial license was obtained meets or exceeds their requirements. Many civil, electrical, mechanical and chemical engineers are licensed PEs.

Licensing is not required for most ceramics engineering professions. However it is recommended to enhance your credentials and make you appealing to more job opportunities. License requirements for ceramics engineers usually include a degree from an American Board for Engineering and Technology accredited engineering program, four years of work experience and successful completion of a state exam.

Few schools offer an undergraduate degree in environmental engineering. Another way to become an environmental engineer is to earn a civil, mechanical, industrial or other traditional engineering degree with an environmental focus. You must pass an Engineer In Training (EIT) exam covering the fundamentals of science and engineering. A few years after you've started your career, you also must pass an exam covering engineering practices. Additional certifications are voluntary.

Not all computer engineers are certified. The deciding factor seems to be if certification is required by the employer. Many companies offer tuition reimbursement or incentives to those who earn certification. Certification is available threw the Institute for Certification of Computing Professionals and the Associate Computing Professionals (ACP) and the CCP, Certified Computer Professional. Certification is considered a sign of industry knowledge. An option if you are interested in software engineering is to pursue commercial certification. These programs are usually run by computer companies that wish to train professionals in working with their products.

Licensing as an industrial engineer is recommended since an increasing number of employers require it. It will also be in you favor to have it when apply for positions. Licensing requirements vary from state to state but in general they require you to have graduated from a accredited school, have four years work experience and have passed the eight hour Fundamental of Engineering Exam (FE). At that point you will be an Engineer in Training (EIT). Once you have fulfilled all the licensing requirements you receive the designation of Professional Engineer (PE)

For Packaging Engineers, the Institute of Packaging Professionals, a professional society offers two levels of certification: Certified Professional in Training (CPIT) and Certified Packaging Professional (CPP). The CPIT is available to college students, recent graduates and professionals who have less than six years of experience in the field. Requirements for this certification include passing a multiple choice test and an essay test. The CPP can be earned by those with at least six years of experience in the field. In addition to the experience requirement, candidates must fulfill two other qualifications from the following: presenting a resume of activities, writing a professional paper or holding a patent. Although certification is not required it is a good idea to obtain it to show that you have mastered specified requirements and have reached a certain level of expertise.

Most plastics companies do not require a bachelor's degree in plastic engineering. Companies that design propriety parts usually require a bachelor's or advance degree in mechanical engineering. The field of plastics engineering overall is still a field in which people and proper experience are scare and experience is a key factor in qualifying you for an engineering position. To obtain a bachelor's degree in plastics engineering you should contact the Society of Plastics Industry (SPI) or the society of Plastic Engineers(SPE) for information about four year programs. National certification is not required. Both SPE and SPI have established voluntary certification programs.

Although there are no licensing or certification requirements designed specifically for quality control engineers some need to meet special requirements that apply only within their industry. Most Quality Control Engineers come out of the industries from which they are employed. Many quality control engineers pursue voluntary certification from professional organizations to indicate that they have achieved a certain level of expertise. The American society for Quality offers certifications including Quality Engineer Certification (CQE). Requirements include having a certain amount of work experience, having proof of professionalism such as being a licensed Professional Engineer and passing a written examination. Many employers value this certification and take it into consideration when making new hirer or giving promotions.

The Institute of Transportation Engineers (ITE) offers certification as a Professional Traffic Operations Engineer (PTOE). To become certified you must have at least four years of professional practice in traffic operations engineering, hold a valid license to practice civil, mechanical, electrical, or general professional engineering and pass an examination.

Larger companies may offer formal classroom or seminar training. Refer to Chapter 4

10. *Always perform your work in a legal and ethical manner*
Engineering is an important learned profession. As members of this profession, engineers are expected to exhibit the highest standards of honesty and integrity. Engineering has a direct and vital impact on the quality of life for all people. Accordingly, the services provided by engineers require honesty, impartiality, fairness, and equity, and must be dedicated to the protection of the public health, safety, and welfare. Engineers must perform under a standard of professional behavior that requires adherence to the highest principles of ethical conduct.(Refer to Chap 3)

Conclusion

Now that you have examined the Ten Commandments of Engineering you now have a better understanding of your goals and reasons for wanting to be an engineer and a better understanding of what you want to do and why. You now have the Ten Commandments to refer to and to guide you into a successful career in engineering.

We have discussed ideas on how to find the right type and position for you. You are now aware of what companies, salary and position would be best for you and how to attain them. You have been introduced to the organization of a typical company. You were shown how to find a comfortable position within the structure. Every type of company has its own goals and policies and you now know how to find a company that you can relate to. Personalities of engineers have been depicted and are familiar to you now. Knowing how to relate to them is also now easy.

We have reviewed education and licensing and you now know how to get the right foundation for what engineering field you want to go into. Team work and meetings were described and you now know how to handle yourself in these situations for success.

The Ten Commandments of engineering gave you ten steps to follow in order to succeed. You have learned the most important factors for success in working in the engineering profession. You can become an engineer. Now you have the knowledge.

Bibliography

Celeste Baine, Is There an Engineer Inside You? 2002
Nicholas Basta, Opportunities In Engineering Careers, 2002
Joyce Hadley Copeland, Where The Jobs Are, 2000
Harvey Kaye, Inside The Technical Consulting Business, 1994
Donna Dunning, What's Your Type of Career?, 2001
Ferguson, Careers in Focus Engineering, 2003
Ferguson, Ferguson's, Guide to Apprenticeships Programs,1998
Geraldine O. Garner, Great Jobs for Engineering Majors, 1995
America's Top 300 Jobs